UNIFORMS OF THE SEVEN YEARS WAR
1756–1763

By the same author

Uniforms of the American Revolution

UNIFORMS OF THE SEVEN YEARS WAR
1756–1763
in colour

JOHN MOLLO

Illustrated by Malcolm McGregor

BLANDFORD PRESS
POOLE DORSET

First published in 1977
by Blandford Press Ltd,
Link House, West Street, Poole, Dorset BH15 1LL

ISBN 0 7137 0822 0

Typesetting by Woolaston Parker, Leicester
Printed & bound by Cox & Wyman, Fakenham

CONTENTS

PREFACE

The Seven Years War, which lasted from 1756–1763, has been described as the first 'world war'. Not only were all the major European powers involved, but the war was fought in North America, the West Indies, and the subcontinent of India, as well as Europe. It saw the rise of Prussia to the rank of a world power, the first intervention of Russia in Central Europe, and the establishment of the British Empire. The successes of Frederick the Great's formidable army were to have considerable influence on the development of European warfare, while, in North America, fierce forest fighting led to the development of light infantry, and to a break with the formal tactics used in Europe. Thus in two main theatres of war developments took place which were to have a lasting importance in military history.

As in our *Uniforms of the American Revolution* I have selected, and Malcolm McGregor has drawn from, wherever possible, contemporary portraits for our figures; and it will be evident that we are much indebted to the work of Menzel for the poses of many of the Prussian soldiers. In general, there is much more information available about the uniforms and equipment of the participants of the Seven Years War than is the case for the American Revolution. The armies were for the most part regular formations whose uniform details were recorded in regulations or in printed lists and establishments. There are however hardly any contemporary representations of the other ranks apart from the paintings of David Morier, in the Royal Collection, and the works of such artists as Watteau, Le Paon, Parrocel, and Joseph Vernet, and little that can give us a true picture of what the eighteenth century soldier really looked like on campaign. We have tried to give a good overall picture, which we hope will lead the enthusiast to delve deeper into the subject.

Confronted with such a wide subject I decided to plan the book around eight specific engagements, of differing importance size and participants. Thus we shall examine a Prussian victory over the French and Imperial troops, another over the Austrians, a Russian victory over the Prussians, four British and allied victories over the French, and a Franco-Canadian victory over the British. Each engagement has a brief description of the action, and an order of battle showing the composition of the opposing forces and their formation at the outset of the action.

The political history leading up to the war has not been touched upon, and for reasons of space the uniforms and operations of the Swedish and Spanish armies, and French and British activity in India, have not been described.

The principle sources consulted have been listed at the end of the book, but I must acknowledge my reliance above all on General Savory's invaluable work on the British and their allies in Germany, and Christopher Duffy's excellent book on the army of Frederick the Great. My thanks go once again to Malcolm McGregor for all his patient and skilful work, to the Directors and staffs of the National Army Museum, London, and the Victoria and Albert Museum, London, to Count Vitetti for patiently answering an apparently endless stream of questions and demands for information, and finally to my father, Eugene Mollo, for help with translation and with the Russian army.

Hungerford 1976. J.M.

I

THE PRUSSIAN ARMY

When Frederick the Great came to the Prussian throne, in 1740, he inherited an army of 83,000 men, painstakingly built up over the years by his father Frederick William I. This extraordinary monarch, the personification of a drill-sergeant in appearance and manner, was to provide the model for all military-minded monarchs for the next century and a half. The stories of his eccentricities are numerous. At a review at Mühlberg he exchanged his father's priceless collection of Chinese porcelain with the Elector of Saxony for a fully-equipped regiment of Dragoons; and his passion for tall men, preferably with broad faces, led him to despatch agents throughout the length and breadth of Europe searching out, and, on occasions, kidnapping suitable recruits for his 'regiment of useless giants'. Frederick William's army, however, was a plaything, not for use but for amusement and show, for 'his whole pleasure and employment was . . . the exercising and reviewing them; he never chose to expose them to the dangers and fatigues of war . . .'.

Before long, however, Frederick II had blooded them in the war of 1740–42 in which he conquered Silesia, and in the second Silesian war of 1744–45, which confirmed him in his conquest and placed Prussia firmly among the foremost nations of Europe.

In 1756, when Frederick opened the Seven Years War by marching into Saxony his army had increased to 84,770 front line infantry, 4,080 supernumerary infantry, 29,446 front line cavalry, 3,180 supernumerary cavalry, 1,700 gunners with 122 heavy and 250 field guns, and 20,000 garrison troops. During the course of the war Prussia lost some 180,000 dead, and 63,000 prisoners to the Austrians alone. One regiment used up no less than 4,474 men, suggesting that it must have been wiped out three times over. Again, to the Austrians only, the Prussians lost no fewer than 204 colours, 52 standards, and 430 cannon as well as numerous mortars, howitzers and other *matérial*. At the end of it all Frederick had achieved little more than the *status quo*, and an enhanced reputation.

The peculiar style of the Prussian army was left virtually unchanged by Frederick, and after both the War of the Austrian Succession and the Seven Years War it was slavishly copied throughout Europe with varying degrees of success. One of the most distinctive features of the

Prussian service was the officer-corps, in theory at least, an exclusive circle of noble warriors, which delighted in its *camaraderie* and 'joy of service'. In practice it was Frederick who to a large extent decided who should or should not be considered a noble, and there were many examples of arbitrary ennoblement. After the Seven Years War so-called 'bourgeois' officers were ruthlessly removed so that by 1786 they amounted to only 700 out of 7,000 officers. Something in the order of a third of the Prussian officers entered the army through the Berlin Cadet Corps, which had been founded in 1717; others came from the Corps of Royal Pages, while the remainder were taken on by colonels of regiments almost in the capacity of private pages. Over one sixth of the general officers, and a varying proportion of the junior officers were of foreign nationality.

After a period of time the newly-joined *Junker* progressed through the ranks of Free-Corporal and Ensign, to the fully commissioned rank of Lieutenant. Further officer training in the regiment consisted either of being sent to Potsdam to see the Guards regiments go through their paces or of being put through their own by visiting Guards officers. Colonels, or *Chefs*, of regiments were advised to establish regimental libraries and to make their young officers write reports and essays, on various military topics, similar to those with which Frederick bombarded his generals. Promotion was strictly by seniority, and was fairly brisk, particularly in wartime, although individuals could be speeded on the way up, or down, through the personal whims of the king. Seriousness, a 'pensive attention to duty', a 'staid, serious appearance', and a lack of humour, were the characteristics which English contemporaries thought chiefly distinguished the dedicated Prussian officer from the more amateur and careless British and French officers.

The rank and file of the Prussian army was made up partly of part-time native conscripts and partly of foreign mercenaries. Under the 'cantonal' system, set up by Frederick William I, between 1727 and 1735, each regiment was assigned a permanent recruiting area around its peacetime garrison. Each year, before the spring reviews, a regimental officer and a civilian officer selected such recruits as were needed. They served for an indefinite period, but there were many classes of people, particularly in trade and industry, who were exempted. The regiments were required to be at full strength during the spring and summer review season only, and for 10 months of the year the native troops were free to return to their civilian trades. their army pay, meanwhile, going into the pocket of their captain. The proportion of native troops to foreigners was something over one third. The foreign element was obtained by recruiting officers who scoured Europe, but especially the minor German states. Occasionally, as in 1756, foreign prisoners were impressed, sometimes in whole regiments, but the cost of recruiting unreliable foreigners was high, and the fact that they had to be watched over all the time to prevent desertion made it difficult to use regular troops as light infantry.

Having joined his regiment the recruit spent a year quartered with an old soldier who guided him through the intricacies of regimental life. He received two Thalers a month, subject to deductions for bread and meat, and, in the infantry, was billeted in a private house in the garrison town. The cavalry were quartered among the surrounding villages where forage was easier to obtain. On campaign seven men shared a tent, dividing among themselves the various fatigues like fetching wood, straw, and rations, and cooking. Bread and meat were issued free on campaign, but everything else had to be bought from the sutler.

Six months of the recruit year were spent on drill. Blind obedience was expected from the rank and file, and was enforced by a wide range of punishments, from everyday blows with fists or stick to various degrees of arrest, minor corporal punishments like the wooden horse, branding, dismissal from the regiment, hanging and shooting. Even the medieval custom of breaking on the wheel was used in Potsdam as late as 1755. The most popular punishment was the spectacular 'running the gauntlet' which often proved fatal to the victim. The fierce discipline of the Prussian army was maintained by a numerous body of NCOs who were usually hardened veterans, often selected personally by the king. It is remarkable, however, that in the cavalry, Seydlitz managed almost entirely to abandon the use of corporal punishment.

The cavalry – cuirassiers, dragoons, and hussars – Frederick inherited from his father was he later claimed 'bad'. Hardly any of the officers knew their business, while the men went in fear of their horses which they hardly ever rode, performing the exercise on foot like infantry. Their worst fault, however, was their slowness, as they showed at Mollwitz in 1741. Frederick immediately set about reforming the cavalry, issuing numerous instructions and orders which may be summed up as requiring greater rapidity in all movements, the threat against the enemy's flanks, the prohibition of the use of firearms in the charge, and the threat to punish any officer who allowed himself to be attacked first with *infamer cassation*. The results were remarkable as the magnificent charge of the Bayreuth Dragoons at Hohenfriedburg only four years later testified. Much of the credit must also go to Frederick's two great cavalry leaders Seydlitz and Zieten. The old three rank formation of the cavalry was retained until 1757, when shortage of manpower forced it to adopt two ranks like the hussars.

With the aid of his drill-master, Prince Leopold of Anhalt-Dessau, Frederick William I set out to create a perfect infantry. The old showy movements of the Dutch exercise designed for the ponderous matchlock were altered and performed in close order; and the rate of fire was increased, by constant practice, and by the introduction of the iron ramrod, from one to five or six shots a minute, although under combat conditions three was the normal number. At the same time methods of marching were improved so that wheels and evolutions could be carried out with speed and accuracy. The final result, inherited and used to great effect by Frederick the Great, was a trained infantryman who obeyed orders promptly and accurately, and who could be relied upon to advance on the enemy, without losing the all-important dressing, loading and firing on the move.

The 'Old Dessauer' trained the Prussian infantry in the complicated tactics of platoon fire, devised by himself and Marlborough during the War of the Spanish Succession, but they proved impracticable in the heat of battle, and by the time of the Seven Years War had been abandoned in favour of massed battalion salvoes. In action it was difficult for the third rank to get in a clear shot, and the custom was introduced whereby the front rank knelt and fired after they had loaded, and the second and third ranks 'locked on' by taking a short side-step, to the right.

When Frederick came to the throne it was the custom for commanders to form their battle line before reconnoitring the enemy. Frederick, however, re-introduced manoeuvre by developing new methods of moving rapidly from column into line, and by mounting mass attacks, in his celebrated 'oblique order', on the unsuspecting flanks of his enemies, after the general action had commenced. The Prussian infantry retained the four rank formation until 1740 when the number was reduced to three. Occasionally, however, shortage of manpower during the Seven Years War forced Frederick to adopt a two rank line. The men were closely packed, elbow to elbow, with the space between the ranks as little as two paces and a foot. The third rank enhanced the firepower of the line and acted as a reserve to the tall men in the front rank. The basic tactical units were the battalion, with a frontage of 150–200 paces, the division, 40–50 paces, and the platoon, 20–25 paces. The army usually marched in open or closed column of platoons, deploying into line either by the parallel march, in which the column wheeled left or right into line to face the enemy, or by a perpendicular approach march when the platoons deployed to left and right by means of the oblique march. This latter deployment could also be achieved *en tiroir*, by which each platoon marched to its left or right by files. All these improvements enabled the Prussians to form up more rapidly than 'any other troops on earth'. One of the most useful devices was the advance of successive battalions in a staggered formation (*échelon*) which was used with great effect at Rossbach and Leuthen.

The weapons of the individual soldier had changed little since the beginning of the eighteenth century. These included the smooth-bore flintlock musket, short of range and inaccurate, but deadly at close quarters, and the long slender bayonet which had long since replaced the pike in most European armies. Battle tactics, firmly bound up in tradition, centred on the defence or reduction of fortified places, in accordance with well-defined procedures, and upon the confrontation of opposing armies in the open field. Units entering the battle were manoeuvred by their officers in tight disciplined formations, company by company, in straight rank and file. The climax came when the two armies were brought face to face in line of battle separated only by a few yards of open ground. In these positions, the infantry would commence firing in volleys, while cavalry attempted to turn the enemy's flank. If the opposition began to waver and show signs of breaking, the commander would order a bayonet charge to win the field and complete the destruction of the enemy force. This kind of fighting required a high level of proficiency and discipline.

During the first half of the eighteenth century artillery had no tactical training to speak of. The men were taught to serve their guns, make up ammunition, and construct batteries; and in Prussia they had an annual course of 14 days target practice. No regulations existed and special orders were needed on each occasion to assemble groups or batteries. The light artillery, with a range of 1,500 yards was allotted to the infantry battalions, and the heavy pieces and howitzers were united in large batteries and employed at suitable points in the battle line. In the attack, after a short period of preparatory fire, the heavy pieces were brought up to case-shot range, about 500 yards, while the battalion guns were manhandled forward, in front of the battalion intervals, to within 200 or 300 yards of the enemy, until the infantry opened fire.

Frederick the Great relied heavily on his artillery to demoralize the enemy so that his cavalry could charge home, and to cover his flank movements instead of skirmishers. Thus his artillery came to be used like infantry, and at Olmutz, in 1758, he ordered 60 pieces to advance on the enemy flank until they were close enough to fire case-shot. He also demanded that his artillery should be able to search out enemy units in 'dead ground' and made much use of howitzers for this purpose. He attached howitzers to the infantry regiments and eventually they amounted to almost one third of his artillery.

In 1731 General v. Linger reduced the number of calibres to four, 3-, 6-, 12-, and 24-pounders. In 1742 a new light 3-pounder was introduced and two were allocated to each regiment. They were served by eight men and drawn by three horses. This piece, and the new 6-pounder introduced in 1755 had 'chambered' breeches, which required a smaller charge than the full calibre breeches. They were found difficult to load, however, and being outclassed by the new Austrian pieces, were replaced, from 1758, with new versions with longer barrels and full calibre breeches. In 1754 the old 12-pounders were replaced with chambered versions, but many were lost at Breslau, in 1757, and were replaced with the heavy old 12-pounder *Brummer* from the fortress of Glogau, which performed with devastating effect at Leuthen. A third 12-pounder, the so-called 'Austrian' pattern was introduced in the winter of 1759/60. In addition there were 18- and 10-pounder howitzers, and the 7-pounder battalion howitzer which appeared in 1758. The 24-pounder gun used at Leuthen was relegated, in 1759, to siege work, in which the 50-, 24- and 10-pounder mortars were also employed.

Following the example given by the king, the Prussian regimental officers tended to look down on the bourgeois technicians who ran the artillery and engineers. They were excluded from polite society, and were refused leave to travel abroad, in case they gave away technical secrets.

Prior to each campaign, or operation, an 'Order of Battle' had to be drawn up by the staff (see *Appendix*). The infantry was usually placed in the centre in two or three lines, some 300 paces apart, with the cavalry on either flank to protect them and to secure a free field of action. The artillery were not usually included in the formal order of battle. The army was further divided into wings each under a separate commander. There was much jealousy between units as to who should take the 'posts of honour' which were on either flank of each line.

An event of great symbolic importance in the development of the Prussian army was the *stilbruch*, or break in style, of 1718, when Frederick William I deliberately abandoned the French modes that prevailed throughout Europe and invented the sober and distinctive Prussian style of military dress. As Fortescue has pointed out, 'not wholly with a light heart', Frederick William was the first great military tailor ever to sit on a throne, and the attention which he paid to such matters produced a result which his contemporaries found strange to say the least. '. . . in order, therefore, that his soldiers might appear to the utmost advantage', wrote an English commentator in a work published in 1759,

> '. . . and not without a view to economy; he caused their coats to be made so very short, that they barely reached halfway down their thighs; and so scanty in the body and sleeves, that they could scarce put them on; their breeches reached scarcely down to their knees; and their hats were so small, as hardly to cover their heads; so that they were forced to have a contrivance to pin them on, for fear of their falling off when they were exercising. Their hair was queued back and powdered, with only one or two small curls on each side of the face; to this was added square toed shoes, with high heels; a long sword, with a broad blade, worn very high upon the hip; and white gaiters, which as well as waistcoat and breeches, (that were generally white also), were continually chalked to keep them clean; for the least speck of dirt on any of them was punished with the utmost ferocity. This appeared the more extraordinary, as, at the time, the prevailing fashion of dress was everywhere totally different, and being added to a sort of uprightness and stiffness in their air and motions, that was peculiar to themselves, made them have very much the appearance of puppets.'

However strange they may have looked at first, it was not long before their prowess in battle led other nations to ape their powder and pigtails, and even civilian fashions were influenced.

By the late 1750s the hat, whose brim had become smaller, was being worn at an angle, set low over the right brow, and tipped up on the left side. The chief Prussian contribution to the hat was the use of woollen pom-poms and tassels of different colours for each regiment.

The Prussian Grenadier cap had a stamped white or yellow metal plate suitably embellished; and there was another type, probably copied from the Saxons, which was worn by artillery bombardiers and later by the Fusilier regiments. By the middle of the eighteenth century the use of heraldic devices, cyphers, and insignia of orders of chivalry, had become general for the decoration of cap plates, gorgets, cartouche boxes, and horse furniture. The Prussian army used the King's cypher, the national emblem of the black eagle with outstretched wings, and the star of the order of the Black Eagle founded in 1701.

Military coats of the eighteenth century were versions of the civilian 'frock', in that they were often double-breasted, with the lapels buttoned back. At the beginning of the century they were cut full with wide skirts, but gradually it became the custom to turn the skirts of the coat back, and to hook, or button, them together to keep them out of the way when on the march, which had the effect of making the low-placed pockets inaccessible. The first major development in the coat came from Prussia, where the poorness of the country, and the rapid increase in the size of the army made strict economy vital, and the amount of cloth in the uniforms was cut to a minimum. By 1729 the Prussian infantry were wearing dark blue coats with red turnbacks, and narrow lapels in the regimental facing colour.

One of the features peculiar to Prussian military dress was the use of tasselled loops on the coat, apparently derived from the *brandenbourgs* on the east European *kaftan*. The men had them in worsted braid, and the officers in gold or silver wire embroidery. The designs of the officers' loops varied considerably, and were one of the first applications of *rococo* – a style much beloved by the King of Prussia, in the design of military uniforms.

The queue was worn throughout the reign of Frederick the Great reaching down to the waist, and was not apparently shortened until 1786. The curls on the temples, called *Schläfenlocken*, *Hammelpfoten*, or *Wrucken*, were arranged differently by regiments, during the Seven Years War. From one curl over each ear, the number increased to four, arranged in two pairs horizontally, or two vertical curls, and one horizontal and one vertical.

Frederick the Great was the author of a new concept in the art of war, namely, that the soldier must not think for himself, but must know his job to perfection, an ideal which could only be found in the principle of individual automatism. Thousands of automata formed the battalions, powerful machines destined to crush everything in their path; and when they had done so the beaten generals of Europe rushed to Potsdam to learn from the King the 'rules of victory'. Superficial observers believed in a new era of military history. They attributed Frederick's success to 'good countenance', 'alignment', 'drill', and the 'Prussian discipline'. It did not occur to them that Frederick's victories were due to his own genius, and that his rules were imposed on him by the peculiar properties of the army which he had inherited from his father. The consuming desire of his admirers, who were legion, to introduce the external aspects of the Prussian discipline into their own armies had in many cases dire results.

2

THE BRITISH ARMY

In 1756 the British army, less than a hundred years old and relatively new in British tradition, was still viewed with dislike by Parliament and country as a potential tool of tyranny and a drain on the treasury. Thus, as always at the end of a war, it had been drastically reduced after the Peace of Aix-la-Chapelle which ended the War of the Austrian Succession in 1748. The establishment of 30,000 men was increased on the outbreak of the Seven Years War to 51,000 and continued to expand throughout the war until it stood in 1763 at 32 cavalry regiments, and 124 regiments of infantry which together with two Highland Corps and the three regiments of Foot Guards made a total of slightly under 150 battalions.

The army was administered by a complex hierarchy of boards, offices, and officials, which contributed largely to the notorious slowness and inefficiency with which nearly all military affairs were transacted. The officers were mainly from the aristocracy or landed classes, many of them the younger sons of prominent families who looked upon a military career as their only path to wealth and fame. There were no academies for the training of aspiring officers and entry into the army was obtained either through personal influence or through money. The would-be officer usually began his career by purchasing an Ensign's commission in a selected regiment, subsequently buying his promotion when a vacancy occurred at a higher rank. In time of war he might use his own money to raise a new regiment, of which, if it was accepted for service, he would become Colonel, an investment which would bring him substantial returns over and above his salary in the form of perquisites derived from regimental income. Since promotion to the highest rank was by a rigid system of seniority, Britain's military leadership was frequently characterized by mediocrity or even incompetence. The idea still lingered that blue blood was fighting blood, and as one military historian has noted,

'. . . the eighteenth-century officer subordinated the military values of expertise, discipline, and responsibility to the aristocratic values of luxury, courage, and individualism. The aristocrat was an amateur at officership;

and it was not for him a vocation with ends and standards of its own, but an incidental attribute of his station in society . . .'

Colonel Mauvillon, one of Prince Ferdinand of Brunswick's staff officers had a low opinion of the British officer. 'Their home customs', he wrote,

> 'incline them to the indulgences of life, and nearly without exception, they all expect to have comfortable means of sleep. This disposes them to acts of neglect of duty which would sound incredible if related. To all this is added a quiet, natural arrogance which tempts them to despise the enemy as well as the danger. It is well known how much these people despise all foreigners; this of itself renders their co-operation with troops of other nations very difficult'.

Some of the worst abuses were done away with by the much maligned Duke of Cumberland who as Commander-in-Chief from 1748–56 changed the whole tone of the War Office, and made vast improvements in the methods of transacting business. Although in many respects a martinet and an unsuccessful field commander, Cumberland was an able and strong administrator. He began his reforms by teaching the officers that they must obey. The number of wheeled vehicles, even for General officers, was reduced, and two only, one for the Colonel, and one for the Sutler, were allowed in each regiment. Officers were always to appear in uniform. He disliked the system of purchase and while his own choices were not always happy (like the ill-fated General Braddock), he was fully alive to the merit of men of the new school like Amherst and Wolfe to whom he gladly gave promotion. The other great reformer of the period William Pitt also backed this new type of officer; he was quite as strict as Cumberland in matters of discipline but less rigid and more humane, thus rendering signal service to the army.

The rank and file were recruited entirely from the lower classes, and all too often from the very dregs of society. There was little in the military life to attract the honest citizen, yet the ranks had to be filled. During the War of the Austrian Succession the concept of limited service was tried out and repeated in 1756–7, when a £3 bounty was offered to every volunteer for three years service. Generally a standard height of 5 ft 4 in. was laid down for recruits and a reward of £2 was paid to the parish for every unemployed man pressed for five years service. But because voluntary enlistment brought few recruits, recourse was constantly had to the courts, magistrates often sparing convicted criminals on condition that they joined the army. According to Fortescue, 'the ranks were filled in great measure by professional criminals, who, passed from regiment to regiment, spreading everywhere the infection of discontent, debauchery, and insubordination'. Nevertheless there were signs that the army, as an institution, was beginning to find acceptance with the nation. In 1740 the nickname 'Thomas Lobster' first appeared, and in the same year printed 'Army Lists' were first published. In 1753 numbers replaced the names of colonels as titles of regiments, all of which gave a greater air of permanence and regularity.

The necessity of turning unruly men of dubious character into well-regulated soldiers was believed to require harsh discipline. Insubordination, neglect of duty, disobedience, desertion, and other crimes were inexorably punished by hanging, shooting, flogging, picketing and other grievous penalties designed more to terrorize the offender's comrades than to reform the culprit himself. No punishment, however severe, seemed capable of stamping out common desertion the last hope of men made desperate by harsh conditions.

The British private soldier served his king and country for 20 shillings per month, little of which he actually saw, thanks to numerous deductions for clothing and necessaries. He lived mainly on a coarse and monotonous diet, notably deficient in fresh meat and vegetables, which was probably not much worse than that of the common people of the day. Food was distributed by commissaries and the usual practice in barracks or camp was for small groups of soldiers to draw their rations as a unit and then cook them as they desired.

In the mid-eighteenth century there were only two combat branches in the British army, the cavalry and infantry, for both the Royal Regiment of Artillery and the loosely organized engineers served under the Board of Ordnance, a government department which provided cannon and munitions of war to the armed forces. If a particular campaign, operation, or expedition required a 'seige train', units of artillery and engineers would be assigned or recruited for the purpose. Once on service the train came under the direct command of the commander-in-chief, to be used as he saw fit.

The cavalry consisted of the Household Cavalry, the Horse Guards and Horse Grenadier Guards, and the line regiments of Horse, and the Dragoons, who were originally mounted infantry. In 1746, for reasons of economy, the 3rd and 4th Troops of Horse Guards were disbanded and the three senior regiments of Horse were converted into Dragoons, with the courtesy title of Dragoon Guards. In 1755 a 'Light Troop' was added to certain regiments of Dragoon Guards and Dragoons, and in 1759 the first regiment of Light Dragoons, the 15th, was raised. During this period the old system of manoeuvre, by which each horse turned on his own ground, was changed to the wheeling of small divisions, but the cavalry still formed in three ranks. The British cavalry had no need to learn the principle of shock action from Frederick the Great having already learned it from Marlborough.

The regiments of infantry had for the most part one battalion consisting of nine companies, one of which, composed of the tallest strongest men, was the 'Grenadier' company whose speciality had originally been the hurling of hand grenades. About 1755 the custom was growing, in imitation of the French, of selecting certain men, usually for their nimbleness, to form a 'picquet company', which was the forerunner of what was eventually to become the Light Company of the battalion. The eight remaining 'battalion' companies left when the Grenadiers were removed, as they frequently were on service, consisted of ordinary infantrymen or 'hatmen' so-called to distinguish them from the Grenadiers who wore caps. The great characteristic of the British infantry during the Seven Years War was the excellence of their fire discipline and the deadly accuracy of their volleys. At Minden they stood motionless until the French cavalry were within ten paces, and at Quebec they swept the French off the field with one devastating volley. Although the Prussian manual and firing exercises found their way into the British army in 1756 the truth is that the British had nothing to learn from the Prussians. Marlborough had already taught them the superiority of controlled platoon-fire long before Frederick the Great was born. All the Prussian school had to teach was the precision of march learned by pendulum and pace-stick and certain undeniable improvements in the manoeuvre of a regiment or battalion.

Close-order drill, an important part of training for combat, was a matter of automated precision that was achieved only through long and frequent practice. Marching was a large part of the soldier's life. Wherever he went, on land, he carried himself on his own two legs, often lugging up to 63 lbs on his back and shoulders. Regiments in the field had to make their own arrangements for auxiliary transport. This was sometimes accomplished through a contractor, who undertook to provide a specified number of horses and waggons; at other times, simply by impressing the necessary carts and animals from the local populace. An army moving overland on campaign often stretched for miles along the road, a long column of marching men directed by mounted officers. With the column was the train of horse-drawn artillery, munition carts, and heavy farm waggons heaped high with casks of salt or pickled meat and other provisions, and officers' chests, and tents.

The use of Pandours and Croats by the Austrians during the War of the Austrian Succession had introduced a new dimension to European warfare, namely the light troops. In the British army their place was initially filled by Highlanders dressed in their native costume. They were armed with short muskets and carbines and at first were drawn up outside the line in the formal order of battle. Amherst and Wolfe, not content with Highlanders alone, in spite of their excellent service, formed bodies of marksmen, often armed with rifles, which marked another step in the progress towards a proper light infantry.

In the use of artillery, especially against stationary targets, there was a great opportunity to employ the science of ballistics. Theoretically gunners were supposed to take into consideration a variety of controlling factors when aiming their cannon; but in actual combat gunnery was largely a matter of eye and estimate. Given the relatively short range of guns used in the field, well-trained artillerymen were able to deliver effective fire on enemy positions, providing significant support for the infantry. Artillery was most important in siege operations to breach walls and to demoralize enemy garrisons. At first the gunners worked with the battalion guns, light 3- or 6-pounders, but in Germany the guns were scientifically concentrated and handled in large masses through the skill of the Count of Schaumberg-Lippe. The variety of guns used was wide: heavy 6-pounders, light and heavy 12-pounders, 24-pounders and a variety of howitzers and mortars. In 1759 the Engineers stood at 61 officers and a strong company of miners borne on the strength of the Royal Artillery. If the French Engineers were the best in Europe, the British were perhaps the worst.

The medical services available to the troops were woefully inadequate, although probably not much worse than that received by most civilians of the time. There was no organized medical service in the army nor were field hospitals regularly maintained. A severe epidemic, or heavy casualties in action, might result in the establishment of a temporary hospital. At other times sick or injured soldiers were kept in their barracks or camp, to recuperate, or else deposited in some civilian household. Epidemic diseases periodically swept through the ranks, filling not only the regimental hospital but the graveyards as well. Conditions of comfort and sanitation in these military hospitals left much to be desired. Regimental surgeons, assisted by their mates, carried the entire burden of medical care in the field, and at times they must have found their work overwhelming. Few of the soldiers could have had much faith in the powers of these men, their instruments, or their remedies.

The accession of George II to the throne of Great Britain led to a gradual adoption of Prussian ideas on military dress. He had only been King for four months when he announced his intention of giving the army fixed clothing. In 1742 the *Representation of the Clothing of His Majesty's Household, and of all the Forces upon the Establishments of Great Britain and Ireland* was published. It was the first book to give illustrations of a man of every regiment of Horse, Foot, and Dragoons, and served as the first illustrated dress regulations in the British army, being to all extents and purposes identical to the Clothing Warrant issued in 1751. The general tenor of these new regulations was to curtail the licence hitherto allowed to colonels regarding the clothing of their regiments, especially in the use of their personal crests and coats-of-arms. In all this George II was assisted by the Duke of Cumberland, who, at about this time, commissioned the Swiss artist David Morier to paint the series of soldiers of various regiments that had served under him in the war. While both the King and Cumberland had a passion for the *minutiae* of military dress, there is no doubt that their strict insistence on cleanliness and smartness, together with correctness and uniformity of dress, were a great help to discipline, and in the building up of the *esprit de corps* which was to become such a vital characteristic of the British army.

One of the most characteristic features of British uniform, indicative of the amount colonels were prepared to lavish on their regiments, was the generous use of worsted braid, usually white, woven with coloured stripes or other designs, for edging the lapels, cuffs, and waistcoats, and for forming button-hole loops.

DISTRIBUTION OF THE LAND FORCES OF GREAT BRITAIN AND IRELAND DURING THE YEAR 1760

In Great Britain under Lord Viscount Ligonier, Commanding-in-Chief:

2 Troops of Horse Guards.
2 Troops of Horse Grenadiers.

5 Regiments of Dragoons.
3 ,, ,, Foot Guards.
23 ,, ,, Foot.

In Ireland under Lieut.-General the Earl of Rothes, Commanding-in-Chief:

2 Regiments of Horse.
8 ,, ,, Dragoons.
17 ,, ,, Foot.

In Jersey, under Colonel Boscawen:

1 Regiment of Foot.

In Gibraltar, under Lieut.-General the Earl of Home, Governor:

6 Regiments of Foot.

In Germany under Lieut.-Colonel the Marquis of Granby, Commanding-in-Chief:

1 Regiment of Horse Guards.
2 Regiments of Horse.
3 ,, ,, Dragoon Guards.
6 ,, ,, Dragoons.
16 ,, ,, Foot.
2 ,, ,, Highlanders.

In North America under Major-General Amherst, Commanding-in-Chief:

21 Regiments of Foot.

In the West Indies:

6 Regiments of Foot.

In Africa:

2 Regiments of Foot.

In the East Indies:

4 Regiments of Foot.

Total:

31 Regiments of Horse and Dragoons.
97 Regiments of Foot.

3

GERMAN AUXILIARY TROOPS FIGHTING FOR PRUSSIA

Of the minor German states which maintained small standing armies in peacetime Hanover, Hessen-Cassel, Brunswick-Wolfenbüttel, and Schaumberg-Lippe, fought on the side of Prussia and Great Britain during the Seven Years War; and Bavaria, Pfalz, Württemberg, Hessen-Darmstadt, and the 'spiritual states' of Trier, Mainz, Cologne, Würzburg and Bamberg, on the French and Austrian side. This division of loyalties was determined by the geographical position of the states, the religion of the princes and their populations, and by the family affiliations of the ruling houses. The strength and readiness of the peacetime armies varied considerably, and was not necessarily related to the size of the state, but rather to the degree of enthusiasm which the prince had for military matters, and to the subsidies which were distributed by the great powers, in return for military help which was not always forthcoming.

These small armies were organized, dressed and equipped, after the fashion of one or other of the two major German states, Prussia and Austria, the northern states mainly following the pattern of the former. The troops were raised, as was normal throughout Europe at the time, by the enlisting of 'volunteers', and when this proved inadequate, by calling out the militia, which was usually of poor standard, and occasionally failed to materialize at all. This was particularly the case when the raising of the militia depended on foreign subsidies, which, in peacetime, were frequently spent on undertakings which had nothing to do with military matters.

The officers and men, at that time, had some fighting experience, as many of the units had taken part in the War of the Austrian Succession, but during the years of peace that followed their skills had grown rusty. In this respect, however, the troops of the northern states were in a better state of readiness than their southern neighbours the 'spiritual states'. The general officers, on the whole, were inexperienced, having had virtually no opportunity in peacetime of exercising with higher formations. Many generals were stuck in out of the way places as *chefs* of regiments, or governors of fortresses, while others had completely removed to diplomatic or court circles; nevertheless the outbreak of hostilities soon brought younger talent to the fore. The north German officer-corps was almost equal to that of Prussia, while that of the southern state was put at a disadvantage by their princes who generally preferred birth to merit as a reason for advancement.

As far as readiness, tactics, drill, movements, field and camp duty, pay and provisions, were concerned there was little difference between any of the German armies. The militia in many cases was organized and trained on the most old-fashioned precepts.

Of the contingents which served on the allied side during the Seven Years War, that of Hanover was the largest, having, in 1757, an establishment of some 29,000 men, consisting of 29 battalions of infantry, 34 squadrons of cavalry, light troops, artillery and engineers. In May 1757, 12 battalions of Hanoverian infantry were summoned to England, which was threatened by a possible French invasion and was extremely short of troops to repel it. Although they were paid by Great Britain these troops were not strictly mercenaries, as they were 'ordered' over by their ruler, George II, as Elector of Hanover, and not requisitioned under the terms of a treaty of subsidy. The Hanoverian troops were old comrades-in-arms of the British, having fought alongside them in nearly every European campaign since Neerwinden in 1692. The rank and file, imbued with a strong Protestant spirit were hardy and brave, and they were commanded by experienced officers. Prince Ferdinand of Brunswick thought them more able to withstand the hardship of war than any of the contingents serving under him, and their proportion of sick was lower. In 1758, two 'new' line battalions were formed, and in 1762, three grenadier battalions. All the grenadier battalions were then made into self-contained units, no longer formed from the grenadier companies of parent battalions. The greatest increase, however, was in the light troops, whose numbers rose year by year. In 1757, they mustered only 915 men; but by 1762 they totalled no less than 4,086. By the end of the war the Hanoverian army had an establishment of 44,252 men, although the actual numbers available for action were 27,219.

The Hanoverian infantry and Gardes du Corps wore red, the rest of the cavalry, Horse and Dragoons, white, the artillery grey and the engineers blue.

In 1755 the Landgrave of Hessen-Cassel entered into a treaty of Defensive Alliance with Great Britain which was invoked early in 1757 when eight battalions were despatched to England. This was not their first visit, 6,000 Hessians having been sent to Scotland in February

1746. During their stay in Perthshire which lasted some four months the Hessian officers and the Scottish gentry found a common language in Latin. In 1756 Frederick the Great had urged Britain to raise and pay for the upkeep of an army of German auxiliaries to safeguard the right flank of the Prussian army. This suggestion was agreed to early in 1757, and the so-called 'Army of Observation' was formed near Wesel. Early in May the Hessian troops returned from England to join this army at Hameln.

The establishment of the Hessian forces, in 1757, was some 16,500 men of which some 12,500 could take the field. They consisted of 12 battalions of infantry, two of grenadiers, four regiments of horse and two of dragoons. In 1758 two companies of foot Jägers were added and in 1759 three companies of artillery. In 1760 the Landgrave reorganized the army forming each line battalion into two at roughly half-strength, a doubtful procedure designed to make the French think the regiments were twice their previous strength. Prince Ferdinand insisted on alloting these half-battalions the tasks previously given to the whole battalion, and this, combined with the fact that the men were fighting on their own soil for much of the time, made the desertion rate very high. Nevertheless the Hessians fought well. By the end of the war the Hessian contingent in British pay numbered 21,756, of which 16,061 were effective.

The Hessian infantry and artillery wore dark blue, the horse white, with regimental facings; and the dragoons light blue with regimental facings.

The troops of Brunswick-Wolfenbüttel, also no strangers to the British, were engaged under a treaty of subsidy made in 1757. The army consisted of eight battalions of infantry, two grenadier battalions, six companies of dragoons, and three companies of artillery providing crews for the battalion guns. The whole totalled some 6,000 men. After a bad start at Hastenbeck, despite the efforts of their general the Hereditary Prince, or *Erbprinz*, and their attempt to slip away, on the orders of the reigning Duke, after the Convention of Klosterzeven, their morale improved drastically. As the war continued the army was increased. In 1759 a hussar regiment, two battalions of grenadiers, and a corps of Jägers were raised. On 14 January 1760 the treaty of subsidy was renewed until the end of the war. In 1761 an auxiliary battalion was raised which passed into Prussian service at the end of the war.

Major-General Lord Pembroke, who arrived in Germany at the head of the newly raised 15th Light Dragoons, had a high opinion of the Brunswick infantry describing them and 'part of the Hessians' as 'noble-bodied, active, handsome, clean, cool, obedient, civil, brave, immoveable, fellows'. By the end of the war the Brunswick army in British pay numbered 9,280 men of which 6,965 were effective.

The Brunswick cavalry were dressed in white with black cuirasses, the hussars in yellow with light blue facings and pelisses, the infantry in dark blue with regimental facings and the Jägers in green with red facings.

The smallest single contingents in the allied army came from Bückeburg and Saxe-Gotha. They were not mercenaries of Britain, but served in accordance with separate treaties made by the rulers of these small states with the Elector of Hanover. The treaty with Bückeburg was signed in 1756, and was renewed until the end of the war, in 1760. It consisted of an artillery contingent, a regiment of infantry and a small corps of 60 mounted and 40 dismounted 'carabineers'. The contingent from Saxe-Gotha consisted of one infantry battalion, clothed in white, which was attached to the Hanoverian army, into which it was incorporated in 1759. At first it was used in a fighting role, but was later employed as an escort for the Hanoverian artillery and pontoon train.

Finally, it should be noted that a small Prussian contingent served throughout the war with the army of Prince Ferdinand. In 1757, six battalions garrisoned at Wesel withdrew when that place was evacuated and after some vicissitudes returned to Prussia. Early in 1758, at the urgent request of Prince Ferdinand, Frederick the Great sent ten squadrons of dragoons and five of hussars to serve with his army. In May, 1760 the dragoons were recalled to Prussia but the hussars stayed on until the end of the war. In 1759 an auxiliary battalion, the *Voluntaires de Prusse*, was raised locally, and later that year the four squadrons of Bauer's Hussars, which Ferdinand had raised on his own initiative were transferred to Prussian pay.

4
AMERICAN PROVINCIAL TROOPS

The defence of the English colonies of North America, against the French and Indians, depended on three main classes of troops; the Militia, the Volunteers, and the Regulars. The charters of the various provinces allowed them to organize their own defence, and as a result a militia system had grown up, based on the Assize of Arms in the home country by which everyone was obliged to possess arms and to serve. For about 100 years between 1640 and 1740 the colonies met the need for defence in different ways. In Massachusetts English veterans were engaged to train the militia, and cannon were purchased. New England raised Trained Bands a designation imported from England. Maryland established a militia system in 1638, and two years later Rhode Island was holding eight training days and two muster days a year. In 1708 South Carolina had two regiments, manning forts and watch stations along her coast.

The militia regiments were organized by counties, each community of which provided a company. There was no fixed establishment because the counties varied so much in population, but the general rule was that all able-bodied men between the ages of 16 and 25 were bound to serve. Company drill was held at more or less irregular periods, and there were about four training and muster days a year of which the former were used for regimental, battalion, and company, drill parades, and the latter for the physical examination of all freemen of the prescribed age limits unless otherwise excused.

The militia was largely untrained, undisciplined and averse to serving far from home. In the colonial economy the impact of a war was felt soon and heavily. The uprooting of civilians to serve, however briefly, as soldiers, created serious problems for their families and communities. Labour was in short supply and agriculture and industry were soon affected, while merchants and other business men had to curtail their operations as clerks, craftsmen and labourers went off to follow the drum. Often in time of war the shortage of labour was aggravated by the decrease of immigration from Europe and the flow of slaves from Africa or the West Indies. War also meant not only the impressment of men and boys but also of provisions, horses, waggons and other goods needed by the armed forces, causing additional distress.

When major military operations were undertaken by a number of states together regiments were raised 'for service' from militia volunteers. They varied in size and length of service from campaign to campaign. Service was usually for one year, but on the renewal of hostilities in the spring, efforts were made to retain the seasoned men, either by paying them through the winter, or by other inducements. During the so-called 'Queen Anne's War' (the War of the Spanish Succession), New England raised two regiments, and Rhode Island and New Hampshire contributed contingents, for an attack on Port Royal in Acadia; and in subsequent operations regiments were raised by Massachusetts, Connecticut, Rhode Island and New Hampshire. The Provincials achieved their greatest success, however, when during the War of the Austrian Succession, a force of some 4,000 troops, raised from Massachusetts, Maine, Connecticut and Rhode Island, besieged and captured the great French fortress of Louisberg. During the Seven Years War Virginian troops were raised for Braddock's expedition to Fort Duquesne, and for William Johnson's expedition to Crown Point. Subsequently regiments were raised by Connecticut, New Jersey, New York, and Pennsylvania. The same William Johnson's opinion was that provincial troops 'acting by themselves' were useless on a campaign. They were, he thought, only fit for expeditions, or an 'excursion of 10 or 20 days continuance', and to cover regular troops through the woods, on their marches either to erect forts, or to 'demolish those erected by the French in their encroachments on His Majesty's Lands'. Some of the Provincials, however, achieved a high standard of discipline, notable among them the New Jersey regiment of five companies raised early in 1755. By 1758 it had risen to a strength of 1,000 complete with Grenadiers. It was, we are told, 'sent into the field in a different manner from those of most other provinces: they are completely cloathed in a handsome uniform and furnished with all necessaries: and they are Muster'd to a Man: in both which articles several of the other provinces are greatly deficient'.

A specialized form of provincial troops were the 'Ranging Companies' often, as in New Hampshire consisting of informal groups organized to hunt down Indians for their scalps, for which they were paid a bounty. Later they became more formally organized into companies named after their various commanders.

Besides the Militia and the Provincials, including Rangers, there were the Volunteer companies raised from time to time in all the colonies, usually in the more thickly populated areas. Although nominally part of the Militia they were, in effect, private military associations, chartered and authorized to bear arms by the respective colonies. They were usually formed from the wealthier members of the militia who had 'substance to undergo the charges and voluntarily enter', who drilled more often, wore uniform, and maintained a unity and discipline quite different from the militia.

During the early period of American history the colonists received little military aid from England. An independent company of British regular troops was sent to New York when the Duke of York was proprietor, and another company was stationed for a short time in Virginia. This same place saw the first considerable body of British troops arrive in America when a combined battalion of the First Foot Guards and Coldstream Guards was sent there during the rebellion of 1677. The Peace of Utrecht, in 1713, gave Newfoundland and Acadia to England, and the garrison in America was increased but only to some 900 men; but, already, in 1710 a British fleet and a regiment of Marines had joined the New England troops in their attack on Port Royal, for the first time seriously interposing in the long strife between the colonists and the French.

The next development was the raising of regiments on the British establishment in the American colonies. In 1739 Oglethorpe's, or the 42nd Regiment, was raised for service in Georgia; and in the same year Spotswood's (43rd) was raised in Virginia, North Carolina, New England, Maryland, New Jersey, and Rhode Island. The companies of this regiment were raised 'for rank', that is, the officers, who were selected by the Colonial Governor, only received their commissions when their companies were mustered at full strength. The companies were supposed thereafter to be paid for by the individual colonies. About this time Lord Halifax suggested the formation of provincial regiments commanded by American officers, experienced in Indian warfare, but dependent on the Crown for pay and promotion thus freeing them from the parsimony and bickering of the Provincial Assemblies. It is possible that the two new regiments raised after the capture of Louisberg, Pepperell's and Shirley's, which were composed of colonists who had taken part in the expedition, were formed with this plan in mind, because although they were composed of 'raw Provincials just raised in the Colonies' they were clothed and maintained as regular British regiments. In 1756 the Royal American Regiment was raised from Pennsylvania Germans to guard the western frontiers, although a number of its officers and men were brought over from Germany and Switzerland.

Thus in the final campaigns in which Britain won Canada from the French, British regulars, sent out at last in large numbers, fought alongside all the traditional types of troops – Militia, Provincials, Volunteers, Rangers, and Regular regiments raised in America.

The dress of the military in America generally followed European fashions. The militia, in general had no uniforms and wore their normal working clothes. By law each was to have a musket, lead, and powder. Regulations in Connecticut, in 1741, called for each militiaman to have either a sword or cutlass, and hunting knives and tomahawks were carried by militia as well as rangers. Powder horns and ball-bags were carried instead of the military cartouche box. On occasions colonial governments issued blankets and watch-coats to their troops. The officers only seem to have worn a semblance of uniform, usually red.

Regiments raised for service were sometimes clothed in a uniform manner, the two regiment's raised for the expedition to Port Royal being dressed in red and blue. Later a uniform coat usually red was sometimes issued, but not the rest of the clothing. The regiments raised after 1739, on the British Establishment, had the cloth for their uniforms sent out from England together with pattern coats as samples. Where uniforms were issued to provincial regiments they were usually red, blue, or green, and occasionally, in the northern provinces 'drab' or brown.

The big change from European fashions came when the frontiersmen and Rangers adopted the style of dress of the Indians who wore the skins of the animals they hunted. The French trappers – the *coureurs du bois* – early on adopted the Indian dress of fringed buckskin shirt, leggings, and moccasins, and gradually the English colonists copied them. The ranger companies were dressed in a more or less uniform fashion based on this hunting dress. In 1758, General Forbes who was to succeed in the capture of Fort Duquesne, where Braddock had so miserably failed, wrote to Colonel Bouquet of the Royal American Regiment, that he had,

> 'been long in your opinion of equiping Numbers of our men like the Savages and I fancy Col: Byrd of Virginia has most of his best people equipt in that manner . . . I was resolved upon getting some of the best people in every Corps to go on a scouting in that stile for as you justly observe, the Shadow may often be taken for the reality. And I must confess in this country, wee must comply and learn the Art of Warr, from Enemy Indians or anything else who have seen the Country and War carried on in it . . .'

Such words from a British regular officer are a measure of how greatly Braddock's defeat, and other events experienced by the army in North America, had affected the once-rigid views of the best British generals.

5

THE FRENCH ARMY

On his death, in 1714, Louis XIV left an army of 40,000 men built up by a succession of ministers, of whom Le Tellier and Louvois were the most celebrated. It was they who gave the French army the individual character which it retained until the end of the Seven Years War. The major part played by the French in the development of warfare, which reached its *apogée* during the seventeenth century, is beyond dispute. The strength of the French armies lay, above all, in the cult of 'shock'. The *furia francese* was fully demonstrated under Condé, Luxembourg, Villars, and Turenne, whose regiments hurled themselves forward with bayonets fixed, impervious to the enemy fire, and frequently forbidden to fire in return. As late as 1750 Marshal Saxe was condemning the 'obstinate practice' of compelling the French front rank to remain motionless, as they had done at Fontenoy, while the enemy had the first shot; it was an article of faith that this display of cold courage struck fear into the hearts of the enemy. To march first against the enemy constituted a 'post of honour' and was a jealously guarded privilege of the *Vieux Corps* of the French army. Such a doctrine, at once offensive and spiritual, accorded pride of place to the morale of the individual. By the eighteenth century, however, the quality of the French army had deteriorated sadly. Timorous chiefs, favourites of favourites, replaced the great soldiers of former years. After the blood-letting of the later years of Louis XIV lassitude replaced enthusiasm, and a desire to limit gratuitous violence, by avoiding battle, replaced the doctrine of shock.

In 1756 the strength of the French army stood at 213,000 men, of which 37,000 were cavalry, 165,000 infantry, 5,400 light troops, and 5,400 artillery and engineers.

While it was not strictly necessary, as it was later, to be of noble birth, the majority of French officers came from the nobility. There was, however, nothing to prevent a capable soldier from buying a company and so rising to the rank of lieutenant-colonel and beyond. In 1734, of ten newly appointed brigadiers, three were *grands seigneurs*, five *petit noblesse de province*, and two of humbler origins. In fact the number of non-noble officers in the French army after 1748, some 4,000, made necessary the Edict of Fontainbleau, of November 1750, by which, in certain cases, they were granted personal or hereditary ennoblement.

The officer candidate entered his regiment at the age of 13–14, as a cadet, becoming a nominal lieutenant at the age of 17. At 20 he was eligible to buy a company, and at the age of 24 or 25, if he was sufficiently rich or well-connected, a regiment. There was little formal education, and, like their British counterparts, young French officers treated their duties lightly. Within a regiment the Colonel sometimes arranged classes for the young officers, but in peacetime the majority were only present with their regiments during the spring and summer. In 1732 an attempt was made to establish a military academy when a school for militia officers was established at Metz, with professors in mathematics, German, dancing, deportment, and skill-at-arms. It was disbanded after the War of Polish Succession, through lack of funds, but its pupils proved so superior to the general run of young officers that they were given commissions in regular regiments, instead of the Militia, for which they had been intended. The *École Militaire*, with 500 cadets, was formed in 1751, moving in 1753 to Vincennes, and in 1756 to Paris.

When on campaign French officers lived in considerable luxury. At the siege of Prague, in 1742, they were attended by no less than 2,500 valets. Within the towns of tents, which formed the French camps, could be found cooks, servants, purveyors, and merchants of all kinds, even troupes of travelling acrobats and performing players. In the French army, according to one contemporary,

> '. . . there was no discipline, no subordination, no order on the march, in the camp, or even in the battlefield. The very subalterns had their mistresses with them, and officers often left their men to accompany them on their march in their carriages. Everything that could contribute to the luxury of the officers was found in the French camp . . . At one time there were twelve thousand waggons accompanying Soubise's army which belonged to sutlers and shopkeepers, though the army was not fifty thousand strong . . . Balls were given in camp, and officers left their posts to dance the minuet. They laughed at the orders of their leaders and only obeyed when it suited them.'

It was hardly surprising that, weighed down with so much baggage, the army only averaged eight miles a day, and that operations, which consisted mainly of sieges, marches, and countermarches, were extremely slow.

The method of recruiting the rank and file resulted in a rag-bag of soldiers, who lacked discipline, and who lost no opportunity to plunder and desert. The French General Saint-Germain said of the troops he commanded at Rossbach that they were 'a band of thieves and assassins, who took to their heels at the first shot, . . . Never has there been their equal. The King has the worst infantry under the heavens, and the most undisciplined, and there is no longer any way of serving with such troops'.

As in Britain the system was one of 'voluntary' enlistment. Recruiting parties scoured the towns and villages, employing the most outrageous ruses to persuade the unsuspecting into signing their lives away for six years, in return for a bounty of 20 *livres*. There was an age limit of 16–50, and vagabonds, escaped convicts, militiamen, coastguards, and seamen, were exempt. Frenchmen could not serve in foreign regiments, and foreigners had to serve in the regiments of their own nationality. Only the populace of Alsace had the right to choose whether to serve in a French, German, or Swiss regiment. After six years, in time of peace, the soldier was released, although, if he wished to, he could renew his engagement with the same company in which he had served. In wartime he could be kept on, receiving a further bounty of ten *livres* for each three years served. A soldier promoted to a higher rank automatically prolonged his service without any additional bounty.

Discipline in the French army was not oppressive. The legal punishments included fines, imprisonment and death, and only in the foreign regiments was corporal punishment used. His modest pay, subject, as in most armies, to various compulsory deductions, condemned the private soldier to a life of miserable poverty, although, in peacetime, he was permitted to eke out his wages by taking work in his garrison town, returning to his quarters each night. In case of sickness he was cared for in one of the 85 military hospitals in France, and when he had completed 25 or 30 years service, he could obtain the favour of a place in one of the 200 invalid companies which garrisoned the fortresses and castles in the interior. Those incapable of further service, to the number of 4,000, were housed amid the splendours of the Hôtel Royal des Invalides in Paris, established in the reign of Louis XIV.

In the event of war, the army could be quickly reinforced by calling out the militia for active service. Established by Louvois, during the reign of Louis XIV, the militia was reorganized into battalions, in 1726, thus providing a permanent reserve for the regular infantry regiments. Militiamen, who had to be single men between the ages of 20 and 30, were selected by parish ballots, and were called up for three years service, during which time they were exempted from taxation. In 1756 the grenadier companies of the militia battalions were formed

into regiments of two battalions each, and being *élite* troops formed p…
of the army reserve. During the Seven Years War 15 battalions of militia served with the army of the lower Rhine, and six with that of the upper Rhine, while others guarded lines of communication, or garrisoned fortresses. In all some 104,000 militiamen were called up during the war. The two main drawbacks of the system were the incapacity of the officers, and the unpopularity of obligatory military service in the country.

The French cavalry consisted firstly of the *Maison du Roi*, and the *Gendarmerie de France*, each ten squadrons strong, which constituted the heavy cavalry, and traditionally formed the reserve. The line cavalry consisted of 61 regiments of Horse, traditionally called 'light Cavalry', 17 regiments of Dragoons, and four Hussars. The total strength of the whole cavalry was 233 squadrons, or 37,000 men. In the field the regiments were formed into brigades of six squadrons, bearing the name of the senior regiment.

Even as late as the War of the Austrian Succession the French cavalry had advanced at the walk, or slow trot. Apart from the *Maison du Roi* the regiments were almost wholly the property of individual colonels, whose main object was to ensure the financial success of their ventures. They were not unnaturally averse to risking their regiments, whether in peace or war, and the horses rarely left their stables during the winter months. According to General Warnery, the French cavalry was very much neglected in peacetime, 'a part of them dismounted, which is not the method to improve them'. The French cavalry had lost all the enthusiasm and shock they had posessed in earlier years, and although new regulations of 1755 laid down measures to rectify the situation, there was insufficient time for improvement as the defeats they suffered at Rossbach, Minden, and Warburg, showed.

The French infantry consisted of 10 battalions of French and Swiss Foot Guards, four battalions of *Grenadiers de France*, 80 French, 10 Swiss. 14 German, six Irish, two Scottish, two Italian, two Walloon, and two Lorraine regiments, totalling 245 battalions, of which 63 were foreign, or 165,000 men. Owing to their low strength, the infantry, like the cavalry were formed into brigades, usually of four battalions, which bore the title of the senior regiment. There were, in addition, a number of corps of 'light troops' first raised during the War of the Austrian Succession.

The infantry had, for some time, suffered from a feeling of inferiority, which was not wholly deserved. At Fontenoy, Raucoux, and Lauffeld the casualties they had succeeded in inflicting were comparable to those which they had themselves suffered, yet both officers and men had been demoralized by the regular volleys of the enemy, and by the feeling that they were confronting a superior discipline and solidarity. For this reason the French concentrated on the improvement and quickening of their fire discipline, at the cost of developing their field movements, which were slow and clumsy. In 1754 the Prussian *pas cadencé* and close-order formations were adopted, and in the following year new regulations specified the formation of the line by passing successive platoons to either flank, by files (the Prussian system known as *en tiroir*), but none of these improvements could be perfected in the short time remaining before the outbreak of war.

For years, too, French military circles had been obsessed by the controversy surrounding the respective merits of line (*l'ordre mincé*) and column (*l'ordre profond*). The advocates of the latter, which they considered to be the natural formation for Frenchmen, wished to give up musketry fire altogether, and to rely on cold steel, and the shock action of a solid column breaking a thin line. They had sufficient influence to get some of their ideas included in the new regulations, although they had not been proved in combat. Thus the regulations of 1755 allowed the use of the normal line, three or four ranks deep, the double line, six or eight ranks deep, and the close column of platoons and divisions, and all were to be put to the test during the coming war.

The varieties of drill in use at the time is exemplified by the visit of Marshal Saxe to Paris, in 1750, to make recommendations to the War Minister, d'Argenson, on a few sample systems. The *Gardes Françaises*, and the regiments of Beauvoisis, Maillebois, Broglie, and Alsace all demonstrated their drills before him, and he found that, while the exercises performed by the first two were 'impressive', the new-fangled methods of the last two, both versions of the Prussian, were to be commended.

The preoccupation with massed fire had led, by the end of the seventeenth century, to the disappearance of the *tirailleur*, or sharpshooter, employed with such effect by Turenne and his predecessors. A number of military men, including Marshal Saxe, considered that the aimed fire of *tirailleurs* was the only effective fire, and that its material results could easily compensate for the greater morale effect of massed volleys. There were always parts of a battlefield, woods, copses, hedges, and buildings, where they could be placed with advantage; and they could be used to occupy important posts speedily and with little loss, as well as covering the movements of an army and giving it time to deploy undisturbed. They had been used for such purposes, to a limited extent, during the War of the Austrian Succession, and between 1748 and 1755 had become a regular feature of the various camps of exercise. At that held in 1754, all manoeuvres were carried out with *tirailleurs*, and combined formations of grenadiers and 'picquets', together with troops formed in six ranks, were tried out. During the Seven Years War de Broglie made considerable use of these 'picquets', who were eventually to become the light infantry companies, or *chasseurs*, of the infantry regiments. He formed one such company in each battalion, and expected the whole of his infantry to fight in 'extended order', if required. He made much use of 'light infantry' at Bergen, in 1759, and in 1760, using them as trained troops as opposed to an irregular rabble, like the Austrian light troops.

The French artillery, put on a proper military footing at the end of the seventeenth century, had retained its separate organization. In its schools of 'cannons, bombs, and saps' its personnel received a sound theoretical and practical education. The 'six calibres of France', retained until the end of the reign of Louis XIII, had given place to a proliferation of differing sizes of weapon, which, apart from mortars and howitzers, numbered 17 by the end of the reign of Louis XIV. In 1732, Lieutenant-General Vallière, the first Inspector-General of Artillery, reduced this number to five, 24-, 16-, 12-, 9- and 4-pounders, the last three of which were used for campaigning purposes; but they were cumbersome pieces, which, once placed in position, were difficult to move, and were consequently frequently lost in action. In 1757, however, a light 4-pounder, *à la Suédoise*, was issued to each battalion of infantry. At about the same time Gribeauval was appointed by d'Argenson to examine the light artillery in the Prussian service, but his labours were cut short by the outbreak of war, during which he was attached to the Austrian service, and at one point was captured by the Prussians. After the war he returned to France and set about reforming the French artillery, eventually introducing the famous pieces which bore his name, and which were used throughout the wars of the Revolution and Empire.

All the regiments of French infantry were uniformly clothed in ample, rather old-fashioned, coats of grey-white cloth, which the major of the battalion received, and the captain of each company had made up by the company tailor and his assistants. Each regiment was

d by the style of the coat, the shape and arrangement of
he cuffs and pocket flaps, as well as by the colour of the
en 1750 and 1760 improvements in the art of dyeing,
doption of coloured collars and waistcoats, but the cut of
the coat remained essentially the same until after the Seven Years War. The *Gardes Françaises*, the *Grenadiers de France*, the German, and Scottish regiments wore blue coats; and the Swiss and Irish regiments red. Of the cavalry the *Maison du Roi* wore blue, with the exception of the *Gardes du Corps* and *Gensdarmes* who wore red; the *Gendarmerie de France* wore red; and of the line cavalry some regiments wore blue, some red, but the bulk wore grey-white coats. The Dragoons wore mostly red, and the hussars a variety of colours.

6

THE AUSTRIAN ARMY

In 1740, the Emperor Charles VI died, leaving the [illegible] to his 23 year old daughter Maria Theresa. Apart fr[illegible] German Hereditary Lands (Austria), she inherited Bo[illegible]avia, Silesia, Hungary, Croatia, Transylvania, various minor enclaves in southern Germany, the Austrian Netherlands (Belgium), the Duchies of Mantua and Milan, and Tuscany, which last came to her through her husband, Francis Stephen of Lorraine. The empire contained some 20 million souls, of varying nationalities and tongues. The power and wealth of Austria had been considerably diminished by a century of continuous warfare, and problems of the Imperial succession. The army, trained in the Franco-Dutch school, had been unable, soon after Maria Theresa's accession, to prevent Frederick the Great seizing the wealthy province of Silesia, and after the War of the Austrian Succession it was taken in hand and reorganized, as part of the general reforms initiated by Count Haugwitz.

In February 1748 a military commission was set up, under the presidency of Prince Charles of Lorraine, to carry out widespread reforms. Notable among its members were Count Daun, and Prince Wenzel v. Leichtenstein. Haugwitz's plans called for an army of 200,000 men, of whom 108,000 were to be provided by Austria and Bohemia, 24,000 by Hungary, and 30,000 each from the Netherlands and Lombardy. The budget for this army, 14 million *Gulden*, was to be raised from the 'Estates' and 'Lands', but was to be administered by central government. The money was to come from a tax levied on the immovable assets of all classes of society, according to the assessments of specially appointed commissioners. Income was to be calculated at five per cent. of the agreed capital value, and of this the nobility were to pay one hundredth, and the peasantry one fiftieth. The suggestion that the nobility should pay any taxes at all was both revolutionary and unpopular; but the case was argued with each individual estate, and those nearest the Prussian border soon agreed to cooperate. The Hungarian magnates, however, who did not pay any taxes at all until well into the nineteenth century, proved the most intractable.

At the beginning of the Seven Years War the Establishment of the Austrian army stood at 44 infantry regiments of 2,408 men each, 10 of 2,000 men, and a battalion of 658, which, with the addition of the

..d of the frontier, or *Grenz* troops, made a total of 142,610. ..valry consisted of 18 regiments of cuirassiers of 818 men each, ..egiments of dragoons of 817 men, and 10 regiments of hussars of ..15 men. With the addition of some 1,000 *Grenz* Hussars, the total for the cavalry was 31,678. 24 German artillery companies, of 2,304 men, and eight Netherlands companies of 768, totalled 3,072, to which could be added two companies of miners and pioneers, totalling 484 men. Thus the peacetime strength of the army, on paper, in 1756, was 177,444 men. In reality, however, Count Daun calculated, at the beginning of 1755, that the average shortfall in the infantry was 200–580 per regiment, and in the cavalry 90–100.

The Austrian army probably gave the able, but impoverished, officer a better chance of rising to high rank than any other in Europe. Nevertheless, many of the reforms, brought in by Haugwitz, were unpopular with the older officers, who found themselves being overtaken by a new breed of young officer, many of whom were foreigners, occasionally even Prussian, engaged for their reputation as drill-masters. Reforms brought in to achieve economies often created dissatisfaction, and longstanding abuses were not entirely eradicated. The sale of commissions, for instance, although officially forbidden, was still practised, and as promotion, within the regiment, was controlled by the Colonel-Proprietor, vacancies were often filled with his favourites. The post of Colonel-Proprietor, bestowed by the Court Military Council (*Hofkriegsrath*) in Vienna, was a privilege, rather than a rank, and when the incumbent died, or resigned, the regiment was re-allocated to one of a flood of applicants. Thus court and family influence remained of importance.

As part of the army reforms, a military academy was established, in 1752, in the old fortress town of Wiener Neustadt, and a new type of officer began to appear, whose professional knowledge, in theory at least, was more important than his lineage. Those nobles, of whom Prince Leichtenstein was typical, who had already made large contributions to the running of the army from their own fortunes, distinguished themselves no less under the new regime. They were joined by professional soldiers like Daun, Lacy, and Laudon, who were to make a good showing when war came again. The reforms of Haugwitz welded the disparate elements of the Austrian army into a coherent whole, which owed its first loyalty to Maria Theresa, who, however, as a woman, could not exercise the close personal control over the army enjoyed by her Prussian adversary. Although she was very popular, and although many would have preferred to see her brother-in-law Prince Charles of Lorraine at the head of military affairs, the will of Count Daun prevailed. The Empress, however, kept in close contact with her army, making frequent visits to its encampments, and earning for herself the title of *Mater Castrorum*, and this, together with the chivalrous devotion with which she was treated by the officers, gave the army a semblance of unity. The *cameraderie* and relationships between Austrian officers was not as good as in the Prussian army, and in the higher ranks there were many divisions and jealousies.

Almost every Land or Province of the Austrian Empire had its own form of serfdom, varying from the mild to the extremely harsh. The bulk of the peasantry lived in poverty, illiterate and often brutish, and it was from this class that the rank and file of the army was recruited, on a quota basis. Until Haugwitz reformed the taxation system, the peasants bore the whole weight of taxation. According to the *Annual Register*, of 1766, the Austrian army was,

> 'chiefly composed out of the class of labourers, vassals of the great lords; they are obedient and patient, and bear without a murmur the greatest hardships; and though their religion does not rise to any degree of enthusiasm, probably for want of being excited by an able leader, yet it keeps them sober and free from vice.'

The system of recruiting, introduced by the *Kaiserpatent* of 4 August 1753, entitled 'For the orderly recruiting and completion of the army', allowed for the establishment of the special 'Completing' Militia, of 24,000 men, to be raised from native Austrians, between the ages of 17 and 40, who were in good health, and capable of bearing arms, which was to be renewed every three years. The militia recruits were to undergo training on Sundays and feast-days, and in camps in November of each year. In addition, each regiment was given a recruiting area, for a certain length of time, which it might scour without fear of competition.

The Austrian army suffered greatly from desertion, and, while the actual punishment was less rigorous than in the Prussian service, death was still the prescribed penalty. In an attempt to improve the situation, the conditions of service were revised, and 6,000 invalids were settled in Banat, Siebenbürg, and Slovenia, while those capable of further active service were formed into invalid companies. In peacetime the army was scattered throughout the towns and villages, as there was insufficient money to build barracks. While attempts were made to quarter an entire unit in one place, they were often far from successful. In Hungary, for instance, the cavalry and hussar regiments were usually quartered in groups of eight men, with sometimes as few as one man alone in a particular place.

The regular cavalry of the Austrian army consisted of cuirassiers, dragoons, and hussars. The cavalry was traditionally strong in order to combat the Turkish cavalry, but according to General Warnery it had not adopted 'that active, vigorous, and decided system of manoeuvring, so essential to the operations of cavalry; and in which its superiority and advantage principally consists'. There were two forms of cavalry tactics, one used against the Turks, and the other against European foes. In the latter case the squadrons advanced, with swords hanging from their wrists, and pistols in hand, in two ranks. On approaching the enemy, the gallop was sounded, the pistols fired at 20 paces, and the attack ridden home at speed. Against the Turks, however, they moved slowly in three ranks, well closed up, in order to be able to halt in good order and fire volleys, 'so great was the terror which these consummate horsemen and their sharp swords had inspired in the Christian ranks'. If the enemy was defeated, part pursued, while the remainder rallied; but this was so little practised that after a charge the troops generally got out of hand. When they tried the same tactics against infantry – charges at the trot, or moderate canter – they generally ended in failure.

The hussars held a position midway between the line cavalry and the irregulars, fighting mainly as 'skirmishers'. Their leading sections generally engaged the enemy in front with fire, while the remainder tried to work their way around the flanks. The irregular cavalry consisted of six Slavonian, one Warasdiner, four Banalist, and four Karlstadt squadrons of frontier, or *Grenz* Hussars, each of 200 men.

The Austrian infantry consisted of German and Hungarian line regiments, and irregular troops, known collectively as 'Croats' and 'Pandours'. The Military Commission decided, after some discussion, to retain the strong regiment of four battalions, each of four companies, and two independent grenadier companies. At the end of the War of the Austrian Succession, each infantry regiment had its own drill for deployment and firing. One might fire in quick time, another in slow, and the same words of command could mean different things in different regiments. After examining the situation, the Military

Commission appointed General Daun to work out a new system for the whole army. The new regulations, of 1749, which reached the army in 1751, laid down that the regiment was to form in four ranks, with the grenadier companies in three ranks on either flank. The Austrian grenadiers acted very much as light infantry skirmishers, covering the advance of the line, halting, firing, loading, and running forward, while the battalions followed in close order. Discipline generally was tightened up, and considerable progress was made in introducing the new system by the formation of large training camps.

The irregular frontier, or *Grenzer* troops, which had proved their worth during the War of the Austrian Succession, were completely re-organized after 1748, although not without a certain amount of resistance from the officers of these units. A regimental establishment was laid down, similar to that of the line, and the *Grenzer* staff was made subordinate to the *Hofkriegsrath* in Vienna, and officers were transferred to regular regiments to learn the new exercise. Under the new organization there were ten regiments of four battalions and two grenadier companies, totalling 4,080 men, and one of six battalions and two grenadier companies totalling 6,000 men. The first ten consisted of two Banal regiments, commanded by the *Bans*, or chiefs, of Croatia, three Slavonian, two Warasdiner, and three Karlstädter, while the eleventh was a Karlstädter Likaner regiment. The establishment of the irregular infantry was 46,800 men, but their complete strength could only be counted on during campaigns against the Turks, for in other theatres only one third could be called out, to be replaced after a year, by a second third, in the same way that only a third were called up for the annual training period of four weeks.

The irregular troops fought, partly on foot and partly mounted, much after the manner of the hussars, but without closed supports, and with less attempt at order. In battle they were chiefly employed on the flanks, and in difficult ground which protected them from the volleys and close-order attacks of the regular infantry. The Austrians had a great advantage over the Prussians in this respect, and made the most of it, but want of discipline and love of plunder, in which the irregular troops excelled, often proved a handicap to their own side, particularly when operating on their own ground.

The Austrian artillery was considerably improved, after 1748, thanks to the efforts of the Director-General Prince Wenzel v. Leichtenstein, who gave it a lightness and manoeuvrability, which it had hitherto conspicuously lacked. To this task the Prince brought extensive tactical and technical knowledge, as well as a large private fortune, which he used to further his aims. In 1756 the artillery was divided into four departments. First came the *Artilleriestab*, which was responsible for the general direction of the artillery brigades and their staffs, as well as matters of recruiting, technical improvements, and supplies of all kinds. Next came the *Feldartilleriekorps* itself, divided into three German, and one Netherlands brigade, each of eight companies, and including in its strength the artillery fusilier regiment and two companies of miners. Each brigade organized a school for NCOs, and there was a Corps school for those who were thought suitable for promotion to officers. The *Feldfzeugamt* was responsible for the storage and maintenance of the artillery *matériel* in peacetime, which was carefully allocated so that it could be concentrated rapidly where required on the outbreak of war. Thus in Vienna was stored enough equipment, including ammunition waggons and accessories, for an army of 60–70,000 men, while in Budweis and Thein were kept 95 3-pounders, and 31 heavier pieces, together with some 500 ammunition waggons of various kinds. Lastly came the *Rosspartei*, or horse department, which on the outbreak of war formed units of 50 men and 100 horses, under the command of an *Unterwagenmeister*, which were responsible for the mobility of the artillery train.

The new artillery pieces, introduced in 1753, were to serve Austria well during the Seven Years War. The field guns consisted of 3-, 6-, and 12-pounders, with 7-, and 10-pounder howitzers. The new 3-pounder regimental piece, of which there were two to each battalion, was particularly effective, hurling three rounds per minute a distance of 500 paces. Two or four pieces, 3- or 6-pounders, stood in the intervals between every second battalion in the first line of the army, with *Flügelbatterien* on each flank. Later in the war, the demand for heavier artillery led to an increase in the number of howitzers used in the field; and, in 1761, a horse artillery detachment was formed, consisting of twelve 3-pounders drawn by four horses, which were attached to the cavalry.

The Engineers were divided into German, Hungarian, Italian, and Netherlands brigades, all under the command of the General-Engineer-Director Prince Charles of Lorraine. The Engineering academy in Vienna (there was a second one in Brussels), however, came under the control of Field-Marshal Count Daun. The *Obrist-Schiffsamt*, which was responsible for the Danube bridges, was also in charge of the two pontoon companies, which were strengthened and re-equipped. Their copper and brass pontoons were replaced with 120 large wooden vessels, designed for fast currents, and 80 made of tinplate, for use in Bohemia and Moravia. 50 wooden pontoons were stored at Pressburg and Ofen, and 20 of wood and metal in Vienna.

In 1747, in an attempt to cut bureaucracy, and to speed up the pace of business, the various departments of the *Hofkriegsrath* were made virtually independent, and under Haugwitz's capable supervision operated far more to the advantage of the army. Clear limitation of responsibility, particularly where the supply of arms and other provisions was concerned, deprived regimental and company commanders of many of the benefits they had hitherto enjoyed, while judicial reforms made them more vulnerable to prosecution. The State's new role, in taking over the financing of the army, necessitated the establishment of provincial commissioners who were responsible to Haugwitz alone, and, as a result, the army was spared the inefficiency and rivalries of petty provincial officials.

The staff of the army was also reorganized, and henceforth every commander-in-chief was assigned a *General-Adjutant*, who prepared the operational orders, and a *General-Quartiermeister*, who regulated the movement of the army, and the construction of camps and positions. He had under his command a *Quartiermeister-Lieutenant*, a Captain of Guides, and a detachment of engineers, as well as the quartermasters and staffs of subordinate units, including drivers, postal officials, and provosts, as well as the *General-Wagenmeister* and his staff.

The arrangements for the sick and wounded of the Austrian army were similar to those in the French army and were somewhat in advance of other countries. In peace or war they were cared for in garrison or field hospitals, and were sent to special convalescent centres before rejoining their units. In action the medical orderlies were posted in the rear of the second line of the army, and brought the wounded back to so-called *Maroderhaüsern*, where they were examined before being sent to the *Feldlazaretten*, and eventually to hospital.

The reformation of the army was accompanied by an attempt to regularize uniforms and weapons, particularly in the irregular regiments where it went hand-in-hand with the process of regimentation. In spite of these efforts, however, there was a lack of detailed orders, and the appearance of a regiment still depended on money, the available stocks of *matérial*, and the taste of the Colonel. Imperial white had replaced the grey-white of natural wool early in the century, but the infantry were now given shorter coats after the Prussian style, and the quality, about which there had been many complaints, was improved. The regulations of 1749 laid down that, when the regiment was under arms, the officers, 'without distinction', were to be clothed in full regimentals, with gaiters, sash, espontoon or fusil. The coats of the officers were not to be turned back. The rank and file were always to be properly dressed, with the hat well cocked, the point turned across the left eye, the hair tied with a black ribbon, gaiters, and with the coat skirts turned back. It was strictly forbidden to embellish uniforms with lace and braid, only field officers being allowed silver lace on their waistcoats. The infantry and cuirassiers wore white, the dragoons, hussars and irregular troops a variety of colours, and the artillery brown.

As to weapons, which were to be kept, at all times, in a good state, with the iron and brass-work well-polished, the infantry received a new, lighter, musket and bayonet, which, however proved clumsy in action and easily breakable, requiring frequent repairs, and eventually a revised model. It is of interest to note that at this time, apparently alone among the participants, the Austrian grenadiers continued to carry hand-grenades.

Eight years of hard work brought about considerable change to the Austrian army, although at the outbreak of war in 1756 there existed a state of unreadiness which slowed down the Austrian mobilization. The transformation, if not perfect, was real enough for the Prussians to admit, grudgingly, after the display of tenacity given by the Austrians at the battle of Lobositz, the first engagement of the war, that their opponents were 'not the same old Austrians at all'.

7
THE RUSSIAN ARMY

At the death of Peter the Great, in 1725, Russia possessed an army of some 200,000 men, which, by 1756 had increased to 331,222. Of this enormous total, 172,240 men belonged to the regular army, 74,548 to the garrison troops, 27,758 to the militia, 12,937 to the artillery and engineers, and 43,739 to the irregular troops. After deducting the Imperial Guard, however, which remained in St Petersburg throughout the war, and taking into account the fact that the army was seldom, if ever, up to full strength, the immense Russian empire had only some 130,000 men with which to undertake offensive operations in central Europe.

Under Peter the Great, the founder of the Russian army, the nobility were obliged to serve the state in return for the lands which they received from the crown. Under Anna Ivanovna the working life of the nobleman was strictly prescribed: the years from 7 to 20 were to be spent in study, and those from 20 to 45 in military, or civil service. The young noble who chose the military life entered the army either from the Corps of Cadets, formed in 1731 by Marshal Munich, or by being enrolled into a particular regiment, and working his way up through the ranks. Those with influence naturally chose the Guards regiments, and the custom, rare under Munich, but increasingly common under the Empress Elizabeth, of enrolling children at birth, meant that when the young nobleman joined his regiment at the age of 16, he was immediately made up to sergeant, in consideration of his length of 'service'. As sergeants in the Guards ranked with sub-lieutenants of the line, the new recruit became an officer overnight, without any other education than he may have received at home. As a result officers increasingly lacked the two qualities demanded of them by Peter the Great, namely, knowledge of their men, and the ability to instruct them in their duties, and discipline became more and more lax, as officers tended to leave matters in the hands of their NCOs. Attempts were made to improve the standards of military education, and although plans to establish a senior military academy came to nothing, officers were encouraged to read and study, and works like Turpin de Cressé's celebrated *l'essai sur l'art de la guerre* were translated and circulated throughout the army.

The bulk of the officers were of Russian nationality, although many nobles of the Baltic provinces, who were not required to serve did so out of choice or financial necessity. They brought to the Russian army the 'German' qualities of energy, tenacity, exactitude, and the advantages of a western outlook; but, being German and protestant, they often failed to understand the virtues and weaknesses of the Russian peasant-soldier.

Recruiting for the regular army was confined to the ten provinces of Greater Russia, and, within those provinces, to the serfs which belonged to the crown, the nobility, monasteries, and other institutions. When an Imperial *Ukase* was issued ordering the raising of so many thousand men, the proprietors selected the recruits to fill their quota. The system was subject to many abuses, and anyone, even the so-called 'free peasants', who could rake up enough money to provide a substitute, or who was prepared to volunteer for more congenial duties, like working for the waggon-train, or as an officer's servant, could obtain exemption. Once in uniform the recruit had little chance of ever seeing his home again, for service was for life; and when he was too old, even for the garrison service, he was given a small pension, and was placed in a monastery, or was given a parcel of land in some far-off border province. As an example of the slowness of this system of obtaining recruits, those raised in 1756 did not reach their regiments until two years later, after having suffered considerable losses, from various causes, *en route*.

The Russian army was entirely national. The rank and file, recruited from the toughest elements of the nation, patient, obedient, and long-suffering, brought to the colours patriotism, and the strong beliefs, tinged with superstition, which had been inculcated in them by their parents and priests. Discipline in the army could be severe, for although the death penalty had been abolished by the Empress Elizabeth, there was still the *knout*, one well-aimed blow of which could break the victim's back, and there was the usual gamut of eighteenth century military punishments. Nevertheless it is an indication of the morale of the Russian soldier that out of 128,000 men serving in 1756, only 185 are recorded as having deserted.

The regular cavalry consisted of the Horse Guards and the *Gardes du Corps*, three regiments of cuirassiers, and 29 regiments of dragoons. In March 1756 the number of cuirassier regiments was increased to five, although only the original three had cuirasses, and three regiments of dragoons were converted to horse grenadiers. In 1757 only 14 regiments, totalling some 7,000 men, were employed on active service, the remainder staying in Russia or guarding the lines of communication.

Since the death of Peter the Great, the cavalry had deteriorated, losing its offensive spirit, and returning to such outmoded practices as firing at the halt, and charging at the trot. New regulations placed emphasis on rapidity of movement, and on shock action with cold steel, but they were not properly introduced until after the war had started, and for a long time the Russian cavalry used both old and new systems. The dragoons, long detached on the Turkish or Tartar borders without any opportunity to train together, were 'as badly trained as their horses', and their officers were renowned for their stupidity. Nevertheless, although the cavalry started the war as the weakest branch of the army, badly mounted and badly trained, they finished up in a good enough state to take on the Prussian cavalry, in equal numbers, and to beat them.

The irregular cavalry consisted of hussars, the only foreign troops in the Russian service, and Cossacks. Two corps of these latter, the Tchougouiev and the Slobodsk, were similar in dress and organization to hussars. The first, raised by Marshal Munich, consisted of 300 Cossacks of Little Russia and 200 baptized Kalmucks, while the second consisted of five regiments, or *Polks*, each of 1,000 men and 2,000 horses, the second horse carrying each man's baggage. Of the other numerous 'armed hosts' of Cossacks, only those of the Don, 15,000 strong, took an active part in the war. Again, during the course of the war, the irregular cavalry improved rapidly, abandoning its numerous spare horses, and acquiring a fearsome reputation for rapid and daring raids.

The infantry consisted of the three Guards regiments, and 41 regiments of the line, most of which bore titles of provinces, towns, or villages. Only 32 were put on a war footing, in 1756, the remaining 12 serving in the interior where they acted as a reservoir of men, to fill the ranks of the service regiments, and Chouvalov's 'Observation Corps' in Poland, being in their turn brought up to strength by drafts of militia men, garrison troops, dismounted dragoons, and even officers' servants. As with the cavalry there was a general decline in the use of shock action, as advocated by Peter the Great, the regulations of 1731 making no mention at all of the Russian soldier's traditional weapon the bayonet, while those of 1754 concerned themselves chiefly with close-order movements and the Prussian drill for firing. Nevertheless, when it came to actual combat, the tenacity and *élan* of the Russian infantry made it equal, and at times, superior, to that of Prussia.

Chouvalov, who was Grand Master of the Artillery, paid much attention to the reform of the artillery and engineers. These arms, he wrote, 'are based on the sciences, and their officers should receive a scientific education'. Schools for both arms were established, and the quality of instruction improved, with special emphasis on mathematics, although tactics, strategy, and military history were also taught. At the same time the artillery itself was the subject of important reforms. In 1757, after the war had started, it received a new howitzer, which outclassed all existing field guns. The 'Chouvalov' howitzer, or 'unicorn', the creation of S. Martinov, was half the weight of its Prussian counterpart, fired solid shot, case, explosive, and incendiary shell, and had a greater range and degree of accuracy when used for indirect fire. It was produced in three calibres, 2-, 1-, and ½-pounder, and gave the Russian artillery a predominant and murderous role in the battles of the Seven Years War. The reputation of these so-called 'secret' howitzers preceded them throughout Europe. Austria begged successfully for one of the new pieces, but Louis XV, when offered one, together with the services of a Russian artillery officer, politely declined.

In 1756 an artillery officer, wearing infantry uniform, was attached to each regiment, to supervize the handling of the six pieces of regimental artillery, consisting of two 2-pounder guns, and four 6-pounder 'mortars'. But, as the pieces were distributed along the front of the regiment, he was unable to supervize them all in person, and the number of guns was reduced, unofficially, to two of each. The field artillery, consisting of 233 pieces in five standard calibres, was formed into brigades of 20 guns; and a heavy artillery unit, of 46 guns, was formed under the title of the Corps of Bombardiers. Thus, from the beginning of the war, Russia was revealed as having the best trained and equipped artillery in Europe. The Artillery Regulations, of 1759, were rich in new ideas, particularly in the fields of indirect fire, movement, and the replenishment of ammunition in the field. They declared that the first duty of the gunner, whatever the circumstances or cost, was to act to the profit of the cavalry and infantry. Chouvalov also created, in 1757, a regiment of engineers, consisting of two companies of sappers, two of miners, and two of workers. He also developed the Russian bridging train, equipping it with pontoons of treated sailcloth, which were a great improvement on the iron or copper pontoons used by the Prussians.

Throughout the campaigns against Frederick the Great, the leaders of the Russian armies never took a decision of importance without consulting a council of war. In addition they were bound to conduct a daily and voluminous correspondence with St Petersburg. As if this were not enough, in 1756, Bestoujev created the 'Council around the Imperial Court', in imitation of the *Hofkriegsrath* in Vienna. 'The conduct and direction of a mass of 55,000 combatants', he wrote, 'the provision of their necessaries, and decisions about their operations, can only be decided by a Council presided over by the Empress'. As a result of this innovation, so contrary to the ideas of Peter the Great, the commander-in-chief became little more than a government representative with the army. Caught between the Council, which deliberated from afar, and the King of Prussia, who operated quickly and decisively close at hand, the commander-in-chief acted like one paralyzed, and it was this handicap which was largely responsible for robbing the Russians of the fruits of their victories.

As for other staff matters, the Russian army, like the French, was plagued with an enormous train, the army of Apraxine, in 1757, having some 6,000 vehicles, which required a third of his manpower to handle; and as for supply and medical services, they were rudimentary in the first case, and non-existent in the second.

The Russian army, with the exception of the Hussars and Cossacks, was dressed very simply in uniforms which were reminiscent of the out-dated styles of the French army, by whom they were clearly influenced. There was, however, a far greater degree of uniformity in the Russian army than in any other at that time, the infantry being all dressed in green with red facings, distinguished only by their territorial badges. The cuirassiers wore buff, the dragoons blue, and the artillery red. In one other thing the Russian army differed from its contemporaries, in that every soldier was issued with a cloak, while the other armies had to rely on blankets, and the occasional 'watch-coat' to protect them from the cold.

8

THE IMPERIAL ARMY

Each of the ten regions, or *Kreis*, of the Holy Roman Empire, was bound to provide a military force for the defence of the Empire should it be threatened with attack. Out of nearly 2,000 component states, however, only some 300 were under obligation to provide men or money. On paper, the Imperial, or *Reichs*, army amounted to 36,000 cavalry and 84,000 infantry, but regional legislation succeeded in reducing these figures considerably, especially of the cavalry.

In 1727 a general staff was established, but communal differences required that there should be an equal number of catholic and protestant generals, even to the extent of having joint commanders-in-chief. Each *Kreis* was supposed to appoint a *Kreisoberst*, but on the rare occasions when the post was filled, lack of money prevented the incumbent from carrying out his two main duties, raising men and providing adequate fortifications.

In September 1757, after the battle of Kolin, when Frederick the Great was in an extremely dangerous position, most of the German princes remembered their obligations to the Imperial crown. Over thirty of them sent contingents to be combined with the French army under Soubise. Some of them, as noted in the following chapter, already had troops serving with either the French or Austrian armies.

The Imperial army as formed in September 1757, amounted to some 29,000 men. Austria which should have provided the largest contingent was only able to spare two cuirassier and two hussar regiments, and then only after the battle of Kolin. Moreover some 6,000 Württembergers hired out to France were unavailable. The command of the army was given to the Prince of Sachsen-Hildburghausen, but his task proved virtually impossible, as every operation and order had to be approved by Vienna. He had under him no less than 25 generals who strenuously opposed the attempts he made to reorganize the army. Of the junior officers a few were good, but the majority, mostly promoted NCOs, were untrained or too old; and in many regions commissions were handed out in return for monetary considerations. The rank and file, recruited on a feudal basis, did no training in peacetime and required much hard work to make them fit for service. Some regiments were so bad that they had to be stiffened by the addition of drafts from the good regiments.

Of the five cavalry regiments the Pfalz Cuirassiers was the only single-contingent regiment, while the Hohenzollern Cuirassiers, on the other hand, 590 men strong, was made up from 61 contingents varying in size from one man to 95. According to Hildburghausen the cavalry 'knew no manoeuvres, and could hardly understand such simple commands as "Fours Right" '.

The 42 infantry battalions and 34 grenadier companies varied in quality. The single-contingent regiments, of which there were eight, were, on the whole, fit for service, while some, like the Hessen- Darmstadt regiment were excellent. The remainder consisted of varying numbers of contingents from six to 42. At first the regiments consisted of anything up to four battalions, or from four/eight companies each, but later they were all reorganized on the Austrian system.

The artillery, consisting of 62 regimental pieces, and 12 reserve pieces, was in a parlous condition. Hildburghausen wished to appoint a chief artillery officer to reorganize the arms, but was refused permission by Vienna. The 3- or 4-pounder regimental pieces, drawn by two or three horses, and their ammunition waggons, were in a terrible state. One badly-trained artillery officer, two *Büchenmeister*, and four *Constabler*, aided by infantrymen from the ranks, served the guns. The reserve artillery consisted of ten 3-pounders, two 12-pounder howitzers, eight ammunition carts, and ten smaller carts. The only exception to this general chaos was the Hessen-Darmstadt regiment, whose artillery was excellent and well-served. Four officers and a conductor comprised the engineers, who were without rank and file labourers, tools, or pontoons.

Each contingent had to provide its own clothing, which was modelled after either the Austrian or Prussian pattern, the older regiments wearing white, and the newer blue. Each contingent was supposed to be dressed in a uniform colour with a distinguishing facing colour, but the smaller states gave their troops only what was available. Generally the uniforms were of very bad quality. The infantry wore hats, grenadier and fusilier caps of Prussian design, and fur grenadiers caps of Austrian style. The cavalry wore white, with the exception of the Württemberg Dragoons, who wore blue; and the artillery were dressed mostly in blue, three Oberrheinisch regiments wearing green.

The weapons were very bad, only some ten per cent. of the available muskets being in working order. The Cologne regiment mustered without any arms at all. Each man had 50 rounds, and carried a bayonet, and sometimes a sword. Some regiments, like the Württembergers, for instance, kept their bayonets permanently fixed in the Prussian manner. The equipment was old and brittle, the cartouche boxes too small, the tents in a miserable state, and some regiments were without blankets for their men.

The supply services matched everything else, and when the army was assembled the arsenals held only enough food for 30,000 men for four weeks. In September, 1757, a Commissary-General was appointed and matters improved slightly. Discipline was bad, and lack of pay and food drove many to desert. The great mistake was made during the war of using the Imperial army in a combatant role, instead of for garrison or lines-of-communication duties. 'They seldom', a modern historian has written, 'if ever, fought well; and at Rossbach did as much to bring about the defeat of the French army as did Frederick the Great's Prussians.'

INFANTRY

		c'tgts	Bns	Gren. Coys	Kreis	Coat	Facings
1	Mainz (Wildenstein)	1	4	2	Rhenish	White	Green
2	Trier	1	2	—	,,	White	Red
3	Cologne (Leib-Regt)	1	1	1	,,	Blue	Red
4	Cologne (Wildenstein)	1	1	1	,,	Blue	White
5	Pfalz (Essern)	1	2	2	,,	Blue	White
6	Cronegk	18	2	2	Franconian	Blue	White
7	Ferentheil	20	2	2	,,	Blue	Red
8	Varell	26	2	1	,,	Blue	Yellow
9	Bayern	1	3	2	Bavarian	Blue	Red
10	Salzburg	12	1	1	,,	White	Red
11	Württemberg (Fuss-Regt)	6	1	2	Swabian	Blue	Yellow
12	Fürstenberg	21	2	2	,,	White	Red
13	Baden-Durlach	27	1	2	,,	Blue	Red
14	Baden-Baden	42	2	2	,,	Blue	White
15	Hessen-Darmstadt	1	2	2	Upper-Rhenish	Blue	White
16	Pfalz-Zweibrüken	33	2	—	,, ,,	Blue	Red
17	Nassau-Weilburg	26	2	2	,, ,,	Blue	White
18	Münster (Elberfeld)	1	1	1	Westphalian	Blue	Red
19	Münster (Nagel)	1	1	1	,,	Blue	Red
20	Paderborn (Mengersen)	1	1	1	,,	Blue	Red
		241	35	29			

	c'tgts	Bns	Gren. Coys	Kreis	Coat	Facings
CAVALRY						
1 Pfalz Cuirassiers	1	3	3	Rhenish	White	Red
2 Bayreuth Cuirassiers	23	5	2	Upper-Rhenish Franconian	White	Red
3 Hohenzollern Cuirassiers	61	4	2	Swabian	White	Red
4 Anspach Dragoons	21	5	2	Franconian	White	Blue
5 Württemberg Dragoons	22	2	2	Swabian	Blue	Black
	128	19	11			

9

GERMAN AUXILIARY TROOPS FIGHTING FOR AUSTRIA

Apart from the dubious assistance of the *Reichs* army Austria and France could call on the services of several of the minor German states. The reasons for their choice of sides, and their military value have already been briefly touched upon in Chapter 3. The largest of the German states, and the one which was destined to bear the brunt of the Prussian attack in 1756 was the Electorate of Saxony. During the years of peace after the 2nd Silesian War the Saxon army deteriorated, largely through the fault of the First Minister, Brühl, whose extravagances left insufficient funds for even the most necessary military requirements. From 1745–6, after the death of the Duke of Sachsen-Weissenfells, Brühl took over command of the army and brought about more reductions and a further worsening of its condition.

At the beginning of the Seven Years War the Saxon army consisted of eight regiments of cavalry, 12 regiments of infantry, and a battalion of grenadiers, a total establishment of 18,617 men. In addition there were five artillery companies, eight companies of pioneers and six garrison units. In 1751–2 new regulations were issued. The infantry was to be formed in three ranks; and the battalion was to be divided into four divisions, eight half-divisions, and 16 platoons. The firing exercise and evolutions were partly Austrian and partly Prussian in inspiration, tending rather more to the latter. The attack was to be carried out in the following manner: the line was to advance in slow time, until within 250–100 paces of the enemy when it was to fire by battalions; it then broke into quick time, halting at 20 paces and again firing battalion volleys; the line was then to hurl itself on the enemy and the front rank, after taking 12 'large paces' was to fire for the last time before lowering its bayonets for the charge. When attacked by the cavalry, infantry were to hold their fire until the enemy were at a distance of 10 paces.

The cavalry was to form in two or three ranks. The squadron was divided into three platoons (*zuge*) each of three sections. Although firing when mounted was practised the charge was to be carried out at the gallop, sword in hand. The regulations for the attack laid down that the line should walk for 20 paces, trot for a further 50–60 paces, and then gallop. The training of large formations took place in special camps, the last of which was held at Ubigau, near Dresden, in 1753.

The infantry were originally foreign mercenaries, but by 1756 they were mostly natives of Saxony. The small size of the companies, and the poor and irregular pay, hampered training, and the assembling of troops in camps of exercise did little to improve the situation. The infantry regiment, according to the establishment of 1753, consisted of two battalions, each of ten musketeer and two grenadier companies, totalling 1,105 officers and men. The grenadiers could be combined to form six grenadier battalions, which with the Kurprinz Grenadier battalion formed two brigades. The *Leibgrenadier-Garde*, of two battalions, totalled 1,684 officers and men.

In addition to the infantry there were the garrisons of Wittenberg, Königstein, Sonnerstein, Stolpen, and Pleissenburg, and the Waldheim Invalid company, in all eight companies totalling 1,150 'semi-invalids'. Possibly even less effective were the four *Kreis* regiments which the sudden collapse of Saxony prevented from joining the *Reichs* Army in 1757. The basic uniform colour of the infantry was white, except for the Leibgrenadiers and the garrison troops who wore red, and the *Kreis* regiments which wore grey.

The cavalry was in an equally bad state, the condition of both men and horses preventing any serious training. The regiment, according to the establishment of 1753, was to consist of four squadrons totalling 514 officers and men, and 393 horses. The *Gardes du Corps*, however, had an establishment of 649. The cavalry wore white except for the *Gardes du Corps* and the Rutowsky Dragoon regiment who wore red.

The Saxon artillery consisted of four battalions of four companies, and the Elector's 'House' company, in all 600 officers and men. The *matérial* consisted of 50 6-pounder regimental pieces, and a reserve of 12 24-pounders, 27 12-pounder, five 6-pounders and four 24-pounder howitzers. The artillery wore green coats with red facings and yellow buttons, and the Pioneers the same but with white buttons. The waistcoats were straw colour, and the breeches white.

The officers were unhappy with their pay and conditions under the arbitrary rule of Brühl, and no less than 378 resigned from the service. At the beginning of the war, however, the command passed to Count Rutowsky under whom the Saxon army put up a surprisingly tough resistance to the Prussians, in 1756, giving Austria time to mobilize. After the disastrous capitulation at Pirna in October 1756 the Saxon army was incorporated by force into the Prussian army, the cavalry being distributed throughout the regiments and the ten infantry regiments taken on entire. Seven of these were disbanded within a year because of heavy desertion, the remaining three becoming the Prussian Fusilier Regiments Nos 54, 55, and 56. In spite of the fact that Saxony was occupied for nearly six years, a large force of Saxons managed to serve with the French under the command of Prince Xavier.

The young Elector Maximilian Joseph III of Bavaria ruled over a state exhausted by years of war, and was forced to observe the strictest economy in military matters. In 1753 he introduced a new establishment for his army which consisted of four regiments of cavalry of 36 squadrons, eight regiments of infantry of 17 battalions, and an artillery brigade, totalling in all 5,100 men. The cavalry was mostly untrained, which was not surprising as for 1,190 men there were only 270 horses. In accordance with a treaty of subsidy with France, Bavaria also had to provide a corps of 10,000 men, but only managed to supply 6,000, who joined the Austrian army in September 1757 serving with it throughout the war. Thus France paid the subsidy but Austria got the troops.

Shortage of manpower forced Bavaria, in 1753, to reform her militia and under the new system, by which every twelfth household had to provide a man, 3,000 young men known as *Landkapitulanten*, were assembled, to serve for three years, but who, after their basic training, were allowed home, and were only called out once a year. The citizens of towns and hamlets were free of this obligation. Generally speaking the exhaustion of Bavaria showed itself in the low quality of both officers and men.

The cuirassier regiment wore white with light blue facings, the two dragoon regiments red with blue and buff facings, and the Hussar Free Corps light blue with red facings. The remainder of the army wore light blue, the infantry with different coloured facings.

The army of the Electrorate of Pfalz consisted, in 1756, of two regiments of cavalry, a squadron of cuirassiers, seven squadrons in all, nine infantry regiments of 18 battalions, and two companies of artillery, in all 12,500 men. In addition, in accordance with a treaty of subsidy signed with Frace in 1756, Pfalz provided a corps of ten battalions totalling 6,000 men. On 1 January 1759 this corps returned home, no longer serving with the French army, in spite of requests for a renewal of the treaty. In 1758 one dragoon regiment served with the Austrian army. The Pfalz officers unless nobles were promoted very slowly, and the purchase of commissions did not help the situation. As the war progressed the various arms received training along Prussian lines. The Pfalz uniforms were white for the cuirassiers, red for the dragoons and blue for the infantry and artillery.

Carl Eugene of Württemberg, who throughout his reign displayed a deep passion for military matters, signed a treaty of subsidy with France, in 1752, by which he was to provide 6,000 men. When they were called for in 1757, he could only produce 3,000, the subsidy having been squandered. He therefore started recruiting the balance, but the mood of the country, being protestant, was against fighting Prussia, and his ruthless methods further estranged the population. He tried to build up his five existing regiments in order to fulfil his obligations, eventually managing to raise ten infantry and three grenadier battalions, which later, under an agreement between France and Austria, were allowed to join the Austrian army. When the French Commissioner reviewed them the infantry regiment Werneck deserted *en masse* at Stuttgart, some of the men returning to the colours after a promise of a general amnesty. All this helped to delay the formation of the units until July 1757, and when, at the beginning of August, Carl Eugene tried to march his troops there were further mutinies. This contingent was still with Soubise in 1758. In 1759 Württemberg signed a new treaty with France by which she was to provide 12,000 men in 15 battalions and 13 squadrons, but because of delays in recruiting, especially the cavalry, they did not join de Broglie's army until October. In 1760, under yet another agreement, they served with the Austrians, at a strength of 17 battalions, 14 squadrons, and some heavy artillery. Throughout four years of war Carl Eugene commanded his troops in person, without any notable success except in the art of extracting subsidies from the large powers, who, while exhausting their treasuries replenished his. He did not participate at all in the campaigns of 1761 and 1762, busying himself with improvements to his army, which he brought home and established in 'cantonments of ease' where he indulged his *soldatenpassion* by re-equipping, clothing and training them in new drills, in an attempt to bring them to what he considered perfection. The Austrian contingent continued to serve until the end of the war.

Two regiments of cavalry wore white with red and black facings, the Horse Grenadiers wore red with black facings and bearskin caps, and the jägers and hussars green with red facings. The infantry and artillery wore blue, the former with various facings and the latter with black.

Hessen-Darmstadt possessed a small, but well-equipped, army, organized and trained on Prussian lines. Its contribution to the Imperial war-effort consisted of a regular *Kreis* contingent, which joined the Imperial army and distinguished itself at Rossbach. The cavalry wore white or buff with red facings, and the infantry blue with red or white facings.

The four remaining states which provided assistance to France and Austria during the Seven Years War were all 'spiritual'. First among them was the Elector Johann Friedrich Carl of Mainz who, as Primate of the Empire, was an inveterate enemy of the King of Prussia and the evangelical church, and who spared no effort to worsen the relations between Prussia and Austria. On the outbreak of war he speedily provided an infantry regiment for Austrian service, which was trained and organized on Austrian lines. The remainder of his army consisted of seven infantry battalions, a company of dragoons, and an artillery and engineer detachment, amounting in all to 5,202 men. From these the *Kreis* infantry regiment Kurmainz was raised for service with the Imperial army, in 1757, again equipped and organized in the Austrian style. The Mainz infantry wore blue, with green, yellow or red facings, and the cavalry blue with red facings.

The Electorate of Trier, another of the spiritual states, possessed only two weak infantry regiments, from which the *Kreis* regiment Kurtrier was formed.

The Elector Clement August of Cologne (*Köln*), Prince of Bavaria enjoyed, in addition, the titles of Prince Bishop of Münster, Bishop of Paderborn, Osnabrück, and Hildesheim, Grandmaster of the German Orders, and Abbot of Stablo and Malmédy. Thus he united under his episcopal crook a large, rich and thickly populated area; but his extravagant habits left him with little time or energy to spend on military matters, and his army, which consisted of seven infantry battalions and two cuirassier regiments, stationed at Münster, and one infantry battalion, at Paderborn, was poorly armed and equipped. The outbreak of the war took the Elector by surprise and he reacted promptly by declaring his neutrality, but in vain, for he had already obtained a subsidy from the French for the provision of 6,000 men. When they were called for in 1757 not one was ready for service, and France had no choice but to provide a new subsidy for 1,800 men, who were recruited directly into the French field army. The contingents sent by Cologne to the *Reichs* army, raised partly from the peacetime units and partly from new recruits, were exceptionally poor and of doubtful usefulness. The units which remained at home, guarding the *Residenz* in Bonn, and other fortified places in Westphalia, lacked almost everything in the way of arms and equipment.

The Prince Bishop of Würzburg, who from 1757 was also Bishop of Bamberg, was another diligent supporter of Austria, and a great enemy of protestant Prussia. In 1756, he obtained a subsidy from Austria for the provision of two regiments of infantry, totalling 4,000 men, one of which he undertook to supply at his own expense. Both these regiments were organized and equipped on Austrian lines. The so-called 'Red' Würzburg regiment, raised from three peacetime units, served with the main Austrian army until the end of the war. After intense recruiting the 'Blue' Würzburg regiment was formed for service with the *Reichs* army. In 1759 one of its two battalions was taken prisoner, the second battalion becoming, in 1761, the third battalion of the Red Regiment. The Dragoon Regiment, which was completely untrained in horsemanship, remained at home, while the small artillery detachment served the regimental pieces and provided a reserve of 12-pounder howitzers for the *Reichs* army.

The permanent *Kreis* contingent of nine companies was swallowed up in the subsidy arrangements and had to be raised from scratch in 1757. The *Kreis* contingent of two cuirassier and two dragoon companies was also formed in peacetime. At the beginning of the war a very strong militia regiment was raised for internal defence and to act as a depot for the field regiments. All this military activity demanded virtual mobilization of the population, to whom it was a heavy burden. They were further embittered by the subsequent heavy contributions and recruiting requirements laid upon them by Prussia.

Rossbach, Prussian Staff Officers

1. Adjutant-General of Cavalry.

2. General of Infantry.

3. Aide-de-Camp of Infantry.

Rossbach, Prussian Cuirassiers

4. Sergeant, Regiment No. 3.

5. Trumpeter, Regiment No. 7.

Rossbach, Prussian Cuirassiers

6. Private, Regiment No. 10.

7. General Officer, Regiment No. 8.

8. Standard-Bearer, Regiment No. 13.

Rossbach, Prussian Cavalry

10. Private, Dragoon Regiment No. 3.

9. Private, Hussar Regiment No. 1.

1. Grenadier, Regiment No. 6.

13. Musketeer, Regiment No. 1.

12. General Officer, Regiment No. 39.

14. Sergeant, Regiment No. 24.

15. Officer, Regiment No. 19.

16. Musketeer, Free Battalion No. 2.

17. Saxon General Officer.

19. French General Officer.

18. French Marshal.

Rossbach, Austrian & Imperial Cavalry

20. Private, Austrian Cuirassier Regiment Brettlach.

21. Private, Swabian Dragoon Regiment Württemberg.

23. Private, Mestre-du-Camp.

22. Officer, La Reine.

Rossbach, French Infantry

24. Private, Royal-Artillerie.

25. Officer, Provence.

26. Fusilier, Piémont.

27. Grenadier, Swiss Regiment Diesbach.

28. Ensign, German Regiment La Marck.

29. Fusilier, German Regiment Royal-Pologne.

Rossbach, Imperial Infantry

30. Grenadier, Upper Rhenish Regiment Hessen-Darmstadt.

31. Officer, Franconian Regiment Varell.

32. Standard-Bearer, Franconian Regimen[t] Ferentheil.

3. Private, Regiment No. 5.

34. Officer, Regiment No. 1.

35. Drummer, Regiment No. 4.

Leuthen, Prussian Infantry

36. Drummer, Regiment No. 10.

37. Standard-Bearer, Regiment No. 13.

38. Oboist, Regiment No.

Leuthen, Prussian Artillery and Light Troops

9. Bombardier, Foot Artillery.

40. Jäger, Free Battalion No. ´.

41. Jäger, Foot Jäger Corps.

Leuthen, Austrian Staff Officers

42. General Officer.

43. Field-Marshal.

44. Engineer Officer.

5. Trumpeter, Regiment O'Donell.

47. Private, Regiment Serbelloni.

46. Officer, Regiment Anhalt-Zerbst.

Leuthen, Austrian Dragoons

48. Private, Regiment Württemberg.

49. Officer, Regiment Zweibrücken.

50. Drummer, Regiment Löwenstein

51. Sergeant, German Regiment Ligne.

52. Grenadier Officer, German Regiment Arberg.

53. Musketeer, German Regiment Merci.

Leuthen, Austrian Infantry

54. Musketeer, German Regiment Charles of Lorraine.

56. Drummer, German Regiment Luzan.

55. Ensign, German Regiment Deutschmeister.

57. Grenadier, Hungarian Regiment Joseph Esterhazy.

58. Corporal, Hungarian Regiment Gyulay.

59. Grenadier, Hungarian Regiment Haller.

Leuthen, Austrian & Allied Troops

60. Gunner, Austrian Artillery.

61. Grenadier, Württemberg Regiment Leib.

62. Musketeer, Bavarian Regiment Leib.

64. Private, Hanoverian Horse Grenadiers.

63. Private, Hessen-Cassel Regiment Leib.

Minden, British Infantry

65. Grenadier, 12th Regiment.

66. Officer, 23rd Royal Welsh Fusiliers.

67. Corporal, 37th Regiment.

Minden, British Infantry

68. Fifer, 25th Regiment.

69. Officer, 51st Regiment.

70. Private, 20th Regiment.

Minden, Hanoverian Infantry

71. Grenadier, Regiment Hardenberg.

72. Officer, Regiment Garde.

73. Sergeant, Regiment Spörcken

74. Grenadier, Brunswick Regiment Imhoff.

76. Musketeer, Schaumburg-Lippe-Bückeburg Regiment.

75. Of⁻icer, Hessen-Cassel Regiment Garde.

77. Gunner, Royal Artillery.

78. Graf v. Schaumburg-Lippe.

79. Gunner, Hanoverian Artillery

80. Gensd'armes Anglais du Roi.

81. Private, Colonel-General.

82. Private, Cuirassiers du Roi.

83. Officer, Royal-Cravattes.

84. Private, Royal-Polog

85. Private,
Dragoon Regiment Schomberg.

86. Private,
Hussar Regiment Royal-Nassau.

Minden, French Infantry

87. Drummer, Rouergue.

88. Grenadier Officer, Auvergne.

89. Fusilier, Regiment du Ro .

90. Fusilier, German Regiment Deux-Ponts.

91. Private, Grenadiers Royaux.

92. Fusilier, Saxon Regiment Lubomirsky.

Kunersdorf, Prussian Cavalry & Artillery

93. Private, Dragoon Regiment No. 6.

95. Gunner, Horse Artillery.

94. General Officer, Hussar Regiment No. 2.

96. Standard-Bearer, Regiment No. 30.

98. Pioneer, Regiment No. 17.

97. Sergeant, Regiment No. 20.

Kunersdorf, Prussian Infantry

99. Fusilier, Regiment No. 35.

100. Officer, Garrison-Regiment No. 5.

101. Sergeant, Regiment No. 41.

102. Pandour, Free Corps v. Kleist.

103. Engineer Officer.

104. Horse Grenadier, Free Corps v. Kleist.

Kunersdorf, Austrian Cavalry

106. Private, Hussar Regiment Kalnocky.

105. Grenadier, Dragoon Regiment Kolowrat.

107. Pandour, Karlstädter Likaner.

109. Musketeer, Free Battalion Loudon.

108. Pandour, Slavonisch-Peterwardeiner.

Kunersdorf, Russian Cavalry

110. Cossack. 111. Cossack Officer.

112. Cornet, Cuirassiers.

113. Dragoon.

Kunersdorf, Russian Cavalry

114. Horse Grenadier.

115. Officer, Hussar Regiment Grouzinski.

116. Sergeant, Grenadier Regiment.

117. Officer, Line Infantry.

118. Drummer, Line Infantry.

Kunersdorf, Russian Troops

119. Musketeer, Observation Corps.

120. Grenadier, Pandour Regiment No. 1.

121. Artilleryman.

122. Grenadier Sergeant, 44th Regiment.

123. Officer, Virginia Militia.

124. Officer, 48th Regiment.

N. America, French & Canadian Troops

125. Private, Quebec Militia.

126. Fusilier, Colonial Infantry.

127. Colonial Artilleryman.

128. Private, Rogers Rangers.

129. Private, Light Infantry.

130. Private, 80th Regiment.

Quebec, British Staff & Highlanders

131 Private, 78th Regiment.

132. General Officer.

133. Grenadier, 42nd Regimen

134. Grenadier, 22nd Regiment.

135. Pioneer, 28th Regiment.

136. Corporal, 15th Regiment.

Quebec, British Troops

137. Ensign, 43rd Regiment.

139. Private,
60th or Royal
American Regimen .

138. Officer, Royal Artillery:

140. Seaman.

142. Lieutenant, Undress.

141. Flag Officer, Undress.

Quebec, French Staff & Artillery

143. Engineer Officer.

144. General Officer.

145. Bombardier, Marine Artillery.

146. Drummer, La Reine.

148. Grenadier Corporal, Languedoc.

147. Officer, Royal Roussillon.

Quebec, French Infantry

149. Fusilier, Guyenne.

150. Ensign, La Sarre.

151. Grenadier, Béarn.

Emsdorf, British 15th Light Dragoons

153. Pr vate.

152. Hornist.

Emsdorf, French Troops

155. Private, Hussar Regiment Berchény.

154. Fusilier, German Infantry Regiment, Royal-Baviére.

156. Private, 1st Dragoon Guards.

157. Officer, Royal Horse Guards.

Warburg, British Cavalry

159. Private, 6th Dragoons.

158. Sergeant, 3rd Dragoon Guards.

160. Private, Light Troop, 11th Dragoons.

161. Officer, Light Troop, 2nd Dragoons.

Warburg, Hanoverian & Hessian Cavalry

162. Private, Hanoverian Regiment Bremer.

163. Private, Hessen-Cassel Dragoons.

164. Private, Luckner's Hussars.

165. Jäger, Freitag's Corps.

Warburg, French Dragoons

166. Private, Regiment Royal. 167. Private, Regiment Thiange.

Warburg, French Infantry

168. Sergeant, Bourbonnois.

169. Ensign, La Couronne.

170. Grenadier, La Tour du Pin.

Warburg, French Chasseurs de Fischer

171. Chasseur à Cheval.

172. Grenadier.

ROSSBACH 5 NOVEMBER 1757

The opening move of the Seven Years War was made by Prussia, when, in August 1756, Frederick the Great marched into Saxony and seized Dresden. Faced with an invading army of 65,000 men the small Saxon army retired to its entrenched camp at Pirna, there to face starvation or surrender. An Austrian army, under Browne, advanced to its relief, but was repulsed at Lobositz, on the Elbe, in October. It retreated in good order and was preparing to renew the attempt when the Saxons were obliged to surrender.

At the beginning of 1757 Frederick the Great's enemies had no less than 430,000 men in the field. While a French army of 110,000, under d'Estrées threatened Hanover, another 30,000, under Soubise, provided the nucleus of a general muster of the German *Kreis*, called out to fight against the enemies of the Empire. Four Austrian armies approached from the south, while 100,000 Russians and 17,000 Swedes hovered on his eastern borders. His only allies, the British, Hanoverians, Brunswickers, and Hessians, under the Duke of Cumberland, awaited the onslaught of the French in Westphalia.

In this situation Frederick took the boldest course and invaded Bohemia; and, after defeating Prince Charles of Lorraine in the Battle of Prague, laid siege to the city in May. In spite of the defeat of their field army, however, the Austrians held out in Prague, and in June an army of 60,000 under Marshal Daun came to their rescue. Frederick with a force of 34,000 attacked Daun, on 14 June, at Kolin, where, as a result of muddled manoeuvring, near-victory was turned into a resounding defeat. The siege of Prague was raised without any interference from the victorious Austrians.

Frederick's enemies were now having considerable success. Prince William, Frederick's brother, defending Silesia with the survivors of Kolin, was hopelessly out manoeuvred by the Austrians. The Duke of Cumberland was defeated by d'Estrées at Hastenbeck, and forced to sign the Convention of Klosterseven by which Hanover was declared neutral. Frederick's only hope now lay in defeating the combined Franco-Reichs army under Soubise, before d'Estrées could come and join it. As he set out through Saxony he was diverted by the news that an Austrian army was marching on Berlin, but at this crucial moment fortune came to his aid. The Russians, misled by a false report of the death of the Empress Elizabeth fell back leaving the Austrians in the lurch; furthermore, both France and Britain repudiated the Convention of Klosterseven, and d'Estrées was unable to come to the aid of Soubise. Frederick resumed his march, reached Leipsig, and after crossing the Saale, on 4 November, encamped near the village of Rossbach, facing the combined army under Soubise and Hildburghausen at Mücheln.

After some preliminary manoeuvring by both sides the battle of Rossbach took place on 5 November. Soubise, planning to seize the heights of Pettstadt and attack the southern end of the Prussian position, started off late on his march, via Pettstadt, to Reichartswerben, and early in the afternoon noticed the Prussians suddenly striking their camp and marching out of sight behind the Janus Hill, apparently retreating. The Imperial army immediately 'rushed recklessly on in an attempt to overtake and crush the enemy. There was no further attempt at reconnaissance, no arrangement for proper deployment'.

Frederick, enjoying a leisurely lunch at Rossbach, paid little attention to this activity, but the young Major-General Seydlitz had already quietly sent a warning order to the army, so that when Frederick eventually realized what was happening, the cavalry and artillery, at least, were ready to move. Placing Seydlitz in command of all the cavalry, Frederick sent him off to the left to head off the advancing enemy. Followed by his staff, Seydlitz trotted along the Janus and Polzen ridge, keeping both the enemy, and his own men, in view, while the 'Green Hussars' (Hussar Regiment No. 1), out in front, kept the enemy from approaching the high ground and spotting what was going on. At the same time 18 heavy guns, under Colonel v. Moeller, moved to the top of the Janus Hill. On reaching the far side of the Polzen Hill Seydlitz formed his squadrons for the attack.

The enemy cavalry advance guard was moving 1,000 ft away from the Polzen Hill, when it was suddenly fired on from the Janus Hill; and had hardly recovered from its surprise, when over the hill swept Seydlitz's cavalry, with drawn swords, 15 squadrons of cuirassiers, and the Meinecke and Czettritz Dragoons, with the Green Hussars moving on the flank. The two allied columns, one 16, the other 17 squadrons deep, were shattered. Only the Austrian Brettlach and Trautmannsdorff Cuirassiers, the Kurpfalz Cuirassiers, and the Württemberg and Anspach Dragoons, managed to advance, under the personal command of Hildburghausen. At the height of the mêlée, when the allies were beginning to break, Seydlitz led on his second line, 18 squadrons, including the Garde du Corps, the Gensdarmes, and the Rochow and

Driesen Cuirassier regiments. After cutting his way through the enemy Seydlitz coolly rallied his men in the hollows near Tagewerben, 13 standards and two kettledrums to their credit.

Meanwhile the Prussian infantry, under Prince Henry, having hurried to the field, went into the attack in *échelon*, 40 paces apart. Some of the French regiments leading the allied infantry, notably that of Piémont, advanced resolutely with the bayonet, but just before they encountered the enemy their resolution failed them and they turned in flight. Once more Seydlitz unleashed his cavalry, and the Gensdarmes and the Gardes du Corps swept forward stirrup to stirrup. The French cavalry brigade La Reine, the Swiss regiments Diesbach and Planta, and the Darmstadt Battalion tried vainly to stem the tide of retreat. The bravery of these few regiments saved the honour of the allied army, while the remainder broke up into swarms of marauding bands.

A modern historian has written that, 'Rossbach stands alone as an example of the superiority of good leadership and high morale over mere weight of numbers, and it is noteworthy as being the first occasion on which a Continental army was inspired to victory by a feeling that can be compared with nationalism in the modern sense'.

Out of 20,000–22,000 men the Prussians lost only 548, while the allies lost a quarter of their strength of 41,000. The Orders of Battle for Rossbach and a map are given at the end of the book.

PLATE DESCRIPTIONS

1. Prussia: Adjutant-General of of Cavalry

The staff of the Prussian army followed the usual eighteenth century pattern, being divided into the Quartermaster-General's, and the Adjutant-General's departments. The former dealt with matters relating to intelligence, routes, orders of battle, camp sites and winter quarters, and the latter with discipline, records, recruiting, the organisation of reviews and manoeuvres, and training. The Quartermaster-General's department in the Prussian service dated back to 1651, when the Great Elector set up a staff on the Swedish model. On Frederick the Great's accession he inherited seven adjutant-generals, who acted in an advisory capacity not unlike that of the modern Military Secretary. Staff officers attached to cavalry formations wore white wtih red facings, adjutant-generals having gold buttons and embroidery, and aides-de-camp (*Flügeladjutanten*) silver.

2. Prussia: King Frederick II

Frederick II (1712–1786), King of Prussia, known as Frederick 'the Great' is shown here in the undress uniform of the 1st Battalion of Infantry-Regiment No. 15, *Bataillon-Garde*. The feather edging to his hat which denotes his General's rank was introduced in 1742. The black breeches and the silver and black Prussian sash worn outside the coat were personal idiosyncracies. He wears the embroidered star of the Order of the Black Eagle on his left breast. A contemporary Englishman wrote of Frederick that,

> 'Either economy, or carelessness, or both, induce him to wear his cloaths as long as decency will permit; indeed, sometimes rather longer. He is accustomed to order his breeches to be mended, and his coat to be pieced under the arms . . . To complete the negligence of his appearance, he takes a great deal of snuff, and lets no small portion of it slip through his thumb and fingers, upon his cloaths.'

3. Prussia: Aide-de-Camp of Infantry

In 1758 Frederick abolished the junior adjutant-generals and increased the number of aides-de-camp to 20. The senior of these were sometimes given command of field units, such as Grenadier battalions, while the junior acted as galloping aides-de-camp. After the Second Silesian War Frederick appointed 12 *Adjutanten bei der Generalen* to enable him to exercise closer control at Brigade or Divisional level. In 1759 he established the first group of officers in the Royal 'suite' who helped aged or ineffective commanders. Staff officers of infantry formations wore blue with red facings, the senior with gold buttons and embroidery and the junior silver.

4. Prussia: Sergeant, Cuirassier-Regiment No. 3. Leibregiment zu pferde

One of the old 'Fehrbellin' regiments, Cuirassier-Regiment No. 3 distinguished itself at Rossbach. The cuirassiers were selected from the best recruits of the Regimental cantons, or recruiting areas, sons of peasants wealthy enough to maintain the horse when their son was sent home on leave. They had to be big men to bear the weight of the iron cuirass and to manage their large mounts. The minimum height of the men was 5 ft 6 in. and the horses 5 ft 3 in. The Holstein breed was particularly in demand as remounts.

5. Prussia: Trumpeter, Cuirassier-Regiment No. 7. (Driesen)

The Cuirassier squadron consisted of 6 officers, 12 NCOs, 2 trumpeters, 2 kettledrummers, 150–160 privates, 1 or 2 farriers, orderlies and staff. There were 740 horses per regiment, with an additional 80 or so for the officers. Cuirassier trumpeters and drummers wore shoulder wings or 'swallow's-nests', and the vestiges of the hanging sleeves worn on the seventeenth century 'cassock' frequently retained in liveries, all laced with special regimental lace which came in two sizes.

6. Prussia: Private, Cuirassier-Regiment No. 10. (Gensd'armes)**

This regiment was stationed at Berlin and was always brigaded with the *Gardes du Corps* on service. The award of 2 stars was made by Frederick when evaluating the services of his regiments during the Seven Years War; and equalled 'very good'. The hats of the cavalry were reinforced with 'secrets' or iron skull caps. The waistcoat (*chemisette*) and jacket (*kollet*) were made of stout buff cloth, which replaced the earlier buff leather coats which had proved difficult to keep clean. The regimental distinctions were found on such items as the lace on the waistcoats the turn backs of the jackets, the edges of the carbine belt, the sabretache and the saddle cloth. The heavy iron breastplate, painted black, was attached by a buff leather strap. Their armament consisted of a carbine, 2 pistols, and a broadsword, or '*pallasch*', with a 40–42 in. blade. In 1756 the grey wool riding coat was replaced by a dark blue cloth cloak.

7. Prussia: Major-General v. Seydlitz

The celebrated cavalry leader is shown here in the uniform of his regiment, Cuirassier-Regiment No. 8**, with the General Officer's feather edging to his hat, and the ribbon of the Order of the Black Eagle. This was usually only awarded to lieutenant-generals, but Seydlitz was a notable exception receiving the order as a junior major-general, for his performance at Rossbach. He carries a *Feldherrnstab*, or baton of command. Born in 1721, Seydlitz was commissioned into a cuirassier regiment in 1740, and took part in the War of the Austrian Succession during which he was taken prisoner. In 1757 he became *chef* of the 8th Cuirassiers, and after distinguishing himself at Prague and Kolin was promoted major-general. Wounded at Rossbach he returned to save the day at Zorndof, but was again wounded at Kunersdorf, this time more seriously. At Freiburg, the last battle of the war, he was in command of a mixed force of infantry and cavalry. After the war he was appointed Inspector-General of the Silesian Cavalry, and in 1767 was made full General of Cavalry. He died on 8 November 1773.

8. Prussia: Standard-bearer, Cuirassier-Regiment No. 13. (Gardes du Corps)

Stationed at Charlottenburg, the *Gardes du Corps* consisted originally of only one squadron, but two more were added in 1756. The regiment distinguished itself at Rossbach Hochkirch and Zorndorf. Being a show regiment permanently stationed at Potsdam, it was highly proficient, and its officers were said to have been so 'well-informed that they could discuss military affairs with any general without the slightest embarrassment'. The *Standartenjunker* in the cuirassiers, like the Free Corporal in the infantry, was an aspiring officer who carried the standard for the nominal *Kornett*. The *Garde du Corps* was the only regiment to wear bright metal cuirasses.

9. Prussia: Private, Hussar Regiment No. 1. (Székely)

The famous 'Green Hussars' who led the van at Rossbach. On his accession Frederick the Great inherited 9 squadrons of hussars, which he had increased by 1745 to 8 regiments of 10 squadrons, totalling some 1,000 to 1,500 men per regiment. The Prussian General Warnery wrote that at first only Hungarians were admitted into the Hussars, but that when the corps were augmented no preference was given to them,' experience having proved that it was not necessary to be born in Hungary to become a good Hussar'. The regiments were assigned no cantons and recruited generally men of a maximum height of 5 ft 6 in. Their small hardy horses were purchased abroad, in southern Poland, Russia, Moldavia, and Wallachia. They were armed with a 41 in. curved sabre, a short carbine and 2 pistols. In 1756 the fur cap shown here was replaced in summer by a felt shako with a wrap around 'wing' (*Flügelmutze*).

10. Prussia: Private, Dragoon-Regiment No. 3. (Meinicke)

The 3rd Dragoons distinguished themselves at Rossbach, but fled under fire at Kunersdorf. Frederick the Great inherited 10 Dragoon regiments which he increased to 12 by 1751. With the exception of the 5th and 6th Regiments, which had 10 squadrons, the others had 5. Dragoons still retained certain features of dress which were vestiges of their earlier role of mounted infantry. Their coats were of infantry cut, their drummers carried side drums, and their carbines had bayonets. The 3rd Dragoons appear to have retained their light blue coats when they left the service of Württemburg, and these probably served as a model for the blue coats introduced generally for the Dragoons in 1745.

11. Prussia: Grenadier, Infantry Regiment No. 6. (Retzow)

Created in 1740 in memory of Frederick William I's celebrated giant grenadiers or 'Long Carls', whose uniforms and weapons it retained intact, the regiment fought at Leuthen, with distinction, at Hochkirch, Leignitz and Torgau. The Grenadier companies of the Prussian infantry regiments, one per battalion, were detached in wartime and combined with the companies of another regiment to form a 4-company grenadier battalion. The Grenadiers of Infantry Regiment No. 6 served during the Seven Years War with those of Infantry Regiment No. 3. Frederick inherited 32 infantry regiments on his accession which were distinguished from regiments created later by their red stocks.

12. Prussia: Field-Marshal Prince Ferdinand of Brunswick-Lüneburg KG

Born in 1721 in Wolfenbüttel, Prince Ferdinand had by the age of eighteen seen action in the Netherlands, France, Italy and Austria. In 1733 his sister married Frederick, and 7 years later he entered the Prussian service becoming *chef* of Infantry-Regiment No. 39. After distinguishing himself in the Second Silesian War he commanded the *Leib-Garde-Bataillon*, being present at Hohenfriedburg and Soor, where he was wounded. He was promoted lieutenant-general in 1750, and at the outbreak of the Seven Years War was in command of the right wing of the Prussian army. Shortly before the battle of Rossbach negotiations had been taking place to secure his services to command the Allied army in Westphalia, to which he succeeded in November 1757. In March 1758 he was promoted to General of Infantry, in September of the same year George II gave him a pension of £2,000 for life, and in December he was made Field-Marshal. In 1766 he entered the Austrian service, and died at Brunswick in 1782. In this portrait, after Tischbein, Prince Ferdinand wears the uniform of his regiment, made up in dark blue velvet, with a British style waitscoat, and the ribbon and star of the Order of the Garter, which he was awarded on 16 October 1759, for his services at Minden.

13. Prussia: Musketeer, Infantry-Regiment No. 1. (Winterfeldt)

This regiment which had an extremely high reputation, distinguished itself at Prague, and Hochkirch, and fought at Rossbach, Leuthen, Leignitz and Torgau. Its Grenadiers served with those of Infantry-Regiment No. 23. The Prussian Infantry regiment consisted of 2 battalions each of which had an administrative and a tactical organization. The first consisted of 5 Musketeer companies and 1 Grenadier company; the second of 4 'divisions' of musketeers, each of 2 'platoons'. The administrative company consisted of 5 officers, 16 NCOs and 114 privates. On service the Prussian infantryman was weighed down with equipment. First came the waistbelt and sword, then the leather cartouche box hanging from a 5 in. wide shoulder belt, then, over the other shoulder an untanned calfskin knapsack 'full of laundry and the like', a linen bread bag, and some item of field equipment, like a tin water flask, a kettle or an axe, and tent pegs. 'Altogether', wrote a private of the Itzenplitz Regiment, 'we were bound five times over by straps passing cross-wise over our chests, and to begin with we all believed we were going to suffocate under the load.' Linen gaiters were worn for parade and taking on service, white for summer, and black, introduced in 1744, for winter. Sometimes false calves were employed to take out unsightly wrinkles. Cloaks for infantry were abolished in 1740 and subsequently the soldiers sometimes shared out the limited supply of 'watchcoats' or wrapped lengths of fur around their middles. Austrian Grenadier caps were especially useful for this purpose.

14. Prussia: Sergeant, Infantry-Regiment No. 24. (Golz)

Recruited in Pomerania and commanded by Field-Marshal Schwerin until his death in 1757, this regiment was famous for its quantity of 'beer-swilling Mecklenburgers'. It fought at Prague, Rossbach, Kay, Kunersdorf, Leignitz, and Torgau. During the war its Grenadiers served with those of Infantry-Regiment No. 34. The 14 NCOs per company were often selected by the King in person; they received extra pay and a 6 lb loaf every 5 days. Their ranks were: *Feldwebel* (sergeant-major), sergeant, corporal, and *Gefreiter* (lance-corporal). They were distinguished by their pike, or *Kürzegewehr* (see also fig. 97) (which had its practical uses, an NCO actually unhorsing an Austrian Cuirassier officer with the hook at Korbitz in 1759), cane, gloves, black and white pom-pom and hat tassels, black and white sword knot, and gold or silver lace on their hats and coats.

15. Prussia: Officer, Infantry-Regiment No. 19. (Margraf Karl)

This regiment fought at Prague, Rossbach, Leuthen, Hochkirch and Kunersdorf, where it was heavily engaged. Their Grenadiers served with those of Infantry-Regiment No. 25. Infantry regiments were commanded by a colonel, occasionally by a lieutenant-colonel, or by one of the majors who normally commanded battalions. The *Hauptmann* commanded the company with the assistance of the Lieutenant and the nominal *Fähnrich*, or Ensign. There were no rank differences in the uniforms except that boots were worn by generals, adjutants, and senior regimental officers who were mounted while others wore gaiters. Black breeches were sometimes worn, and in bad weather a dark blue cloak or greatcoat.

16. Prussia: Musketeer, Free Battalion No. 2. (Mayr)

During the Seven Years War Frederick the Great was driven to raise a number of so-called 'Free battalions' in order to combat the menace of the Austrian Pandours. These brigand-like mercenary bands were recruited from the dregs of mankind. Lieutenant-Colonel Mayr, from Saxony, was given permission to raise his battalion on 18 August 1756. The original battalion was captured at Landeshut, revived and again captured by the Russians in 1761. A second battalion was raised in 1760. Known as the *Zweimalblauen* or 'double blues' from their blue coats and facings, the Free battalions dispensed with tents and kept the rest of their baggage to a minimum. Their artillery was of the light regimental type and was served by men of the battalion. At full strength a Free battalion was between 500–700 men, although they were often down to one third or a quarter strength.

17. Saxony: A General Officer

This portrait of an unidentified Saxon officer carrying a baton of command, may possibly be of Prince Xavier of Saxony, otherwise known as the Comte de Luscace, brother of the Dauphine of France, who, in 1758, commanded 9,000 Saxons in the army of Contades. He wears a typically Saxon uniform with a silver and red sash. Over his left shoulder he wears the ribbon of the Polish Order of the White Eagle, with the badge on his right hip.

18. France: Prince de Soubise, Marshal of France

Charles de Rohan, Prince de Soubise was born in 1715, into one of the oldest families of France, and served as an aide-de-camp to Louis XV during the War of the Austrian Succession. In 1757 he was defeated at Rossbach, but returned to command in Hesse the following year, where he won an action at Lutterburg and was made a Marshal. In 1759, he was in command of the army intended for the invasion of England. In 1761 he was back in Germany where he was defeated at Vellinghausen. 'A timid and indecisive commander . . . said to have owed his high military appointment to the influence of Madame de Pompadour, and (after her death) to Madame de Barry'. He died in 1787, the last of his line. There was no regulation uniform for French General Officers, until 2 September 1775, and in this portrait Soubise wears the elaborate full dress of the *Gensdarmes du Maison du Roi*. On his left breast he wears the Cross of the Order of St Louis.

19. France: Lieutenant-General Francis de Chevert

Born in 1695, Chevert was a professional soldier of humble birth who rose, through sheer merit, to become one of France's toughest and most experienced generals. He took a leading part in the Battle of Hastenbeck which led to the Convention of Klosterseven and the removal of the Duke of Cumberland, and afterwards served under Contades against Prince Ferdinand of Brunswick. Early on in the eighteenth century French general officers adopted a blue and gold campaign dress, which eventually became their regulation uniform. Chevert is wearing such a coat in his portrait, together with the Star of the Order of St Louis on his left breast above that of the Polish Order of the White Eagle. The Badge of the Order of St Louis appears on his left hip.

20. Austria: Private, Cuirassier-Regiment Brettlach

Raised in 1672, this regiment with the Cuirassier-Regiment Trautmanns-dorff particularly distinguished itself at Rossbach. It also fought at Lobositz, Prague, Meiningen, Meissen, Maxen, Strehlen, and Freiberg.

There were 18 cuirassier regiments in the Austrian service, each with an establishment of 1,000 men divided into 6 field squadrons, 1 reserve and 1 'Carabinier' squadron, each of 2 companies. The strength of the field company was 3 officers, 4 NCOs, 1 trumpeter, and 80 privates. The Carabinier squadron, which corresponded to the Grenadiers in the infantry received higher pay. The cuirassiers wore hats with 'secrets', 'Imperial' white coats, and blackened breastplates. They were armed with a long broadsword, and a carbine.

21. The Empire: Private, Swabian Dragoon Regiment Württemberg

One of the few *Kreis* regiments to distinguish itself at Rossbach. The regiments of each *Kreis* were made up of several different contingents sometimes differently dressed and equipped; this regiment contained 22. The Dragoon shown here has a typically Prussian uniform.

22. France: Officer, Cavalry-Regiment La Reine

Raised in 1635, and numbered 14th this was one of the so-called 'Royal Regiments'. In 1756 the French Horse consisted of 61 regiments, 59 of 2 squadrons, 1 of 3 squadrons, and 1, the Carabiniers, of 10 squadrons. They were divided into the regiments of the General Staff, Royal regiments, and the Regiments of the Princes and Gentlemen. A squadron consisted of 4 companies and totalled 8 officers and 140 men. An order of 1 June 1750 simplified the cavalry uniforms considerably. All were to have so-called *Bavaroises Allemandes*, coats with lapels which reached down to the hem, a short buff sleeved waistcoat *à la Prussienne* with small red cuffs, and leather breeches. All hat lace was to be silver. Regiments were distinguished by the colour of their coats and facings, and further by the colour and design of the livery lace used for shoulder straps, cloak 'brandenbourgs', and on the housings, and finally by the shape and design of the buttons. This strict uniformity was unpopular and before long the

Regiments of the Princes and the General Staff were finding ways to get around them, followed closely by the Regiments of the Gentlemen. After Rossbach, where the French cavalry had difficulty rallying after the battle because of the lack of clearly visible distinctions between regiments, the complaints increased and many irregular features crept in. A manuscript of 1761 in the Ministry of War showed most of the cavalry still conforming to the regulations, although many of them had turndown collars and fringed epaulettes. Officers dressed as their men but in better quality materials, and as yet there were no visible rank differences.

23. France: Private, Cavalry-Regiment Mestre-du-Camp Général

Raised in 1635 from former *compagnies d'ordonnance*, and ranking second in seniority, this was one of the three 'General Staff' regiments. The others were the 1st Regiment, the *Régiment Colonel Général*, and the 3rd, *Commissaire-Général*. Each regiment had its own special badge consisting of a trophy of flags. The 1st Regiment 1 blue, 1 red, and 1 white; and the 3rd Regiment 1 blue and 1 white. Cavalry privates were armed with a carbine, two pistols, and a sword which was much criticized for its shortness. They wore iron skull caps or *calottes* and breastplates. After Rossbach the need for cavalry to fight on foot against infantry led to the adoption of bayonets for the carbines and lighter boots.

24. France: Private Royal-Artillerie

Created in 1670 as a guard for the artillery train, the so-called Fusiliers became the Royal-Artillerie in 1693 subsequently ranking 47th in the list of seniority. In February 1720 the artillery was organized into 5 battalions based on Fère, Metz, Strasbourg, Grenôble, and Besançon. Each battalion had a technical school attached to it which turned out qualified gunners. The battalion itself consisted of 5 companies of gunners, 2 of bombardiers, who looked after the howitzers and mortars, and one of sappers. In 1722 the artillery adopted the blue and red uniform shown here, but in 1758 it was slightly modified by the addition of a narrow strip of red cloth down the front of the coat. The artillery were armed and equipped like the infantry, but removed their equipment and coats when serving their pieces. The officers were distinguished by their red stockings.

25. France: Officer, Infantry-Regiment Provence

Raised in 1676 this regiment ranked as the 61st. During the Seven Years War the French had 80 regiments of infantry, 12 of 4 battalions, 2 of 3 battalions, 50 of 2 battalions, and 16 of 1. Each battalion consisted of a grenadier company and 12 fusilier companies. The fusilier companies consisted of 2 officers, 2 sergeants, 3 corporals, 3 lance-corporals, 1 drummer, and 31 privates. The Grenadier company had a sub-lieutenant and 5 grenadiers more. A number of men were detached from the fusilier companies to form a 'picquet' company, which, during the Seven Years War, became permanent with the title of the *Compagnie de Chasseurs*. They do not appear to have had any particular dress distinctions. French officers wore the same uniforms as their men, but made up of better materials. They had no rank badges other than the universal sign of commissioned rank, the gorget. Although shown here with an espontoon, officers tended to carry fusils and bayonets when on service.

26. France: Fusilier, Infantry-Regiment Piémont

Raised in 1569 and ranking as the 4th, this regiment distinguished itself at Rossbach and Bergen. All the French infantry regiments wore a grey-white cloth coat, or *justaucorps*, distinguished regimentally by the cut of the coat, cuffs, and pockets, by the colour and arrangement of buttons, and by the colour of cuffs and lapels if any. Waistcoat, breeches, linen gaiters for summer, and wool for winter completed the basic clothing. After 1757 collars, at first little more than flat pieces of cloth or stoles, lapels, and heart-shaped skirt ornaments became more common, and the cuffs became smaller. The musket was the pattern of 1728 modified in 1746 and 1754. The fusilier's equipment consisted of a cartouche pierced for 19 rounds, below which hung a powder flask. Some regiments, in addition, carried a small pouch on the waistbelt from which hung the bayonet and straight sword. The knapsack, slung over the right shoulder was made of strong waterproofed linen, closed with a cord like a kit bag; a gourd-shaped water flask completed the soldier's personal equipment, but each of the 36 men in the company had to carry, in turn, his share of the 10 tools, 13 items of cooking equipment, 5 tents, a bell-of-arms, and bread for 4 days.

27. France: Grenadier, Swiss Infantry-Regiment Diesbach

Raised in 1690 and numbered 77th, this regiment distinguished itself at Rossbach, Sonderhausen and Korbach. Apart from his national regiments the French king maintained several regiments of foreign mercenaries, mainly Swiss, Irish, and Germans. The Swiss differed from the others in that they served under treaties, or *Capitulations*, made between the King and the Swiss cantons, and sometimes of very long standing, the most recent one being that signed on 9 May 1715. During the Seven Years War there were 20,000 Swiss in French service, all distinguished by their red uniform coats. Otherwise the cut of their uniforms, their arms and equipment were the same as the French infantry. Although many French grenadiers wore the hat, some regiments wore a bearskin cap without a front plate. Other distinguishing marks of the Grenadier were the moustache, the curved sabre, and, in place of the cartouche, an ample leather pouch called a *Grenadière*, which originally contained hand grenades, and to which a hand axe was normally attached.

28. France: Ensign, German Infantry-Regiment La Marck

Raised in 1682 and numbered 6th, this was one of the German regiments in French service, some of which dated back to the days of Louis XIII. The King paid them and furnished them with clothing, arms and ammunition. The chief distinguishing feature of their dress was their lightish blue coats; otherwise they were armed and equipped as the French infantry. By 1749 each French infantry battalion had one white Colonel's Colour and two *Drapeau d'Ordonnance*, 1.78 metres square which varied in design regimentally.

29. France: Fusilier German Infantry-Regiment Royal Pologne

One of the regiments raised on the occasion of the marriage of Marie Leczinska, daughter of Stanislas the deposed King of Poland, to the Dauphin later Louis XV in 1725 and typical of the regiments where the men were German and where the communal interests of the King of France and the German Princes were served. The Fusilier shown here has the blue coat worn by German regiments and the usual French infantry equipment and weapons.

30. The Empire: Grenadier, Upper-Rhenish Infantry-Regiment Hessen-Darmstadt

Another of the few *Kreis* regiments which distinguished itself at Rossbach. Each *Kreis* battalion was formed from several territorial contingents, sometimes variously dressed, and consisted of 1 Grenadier and 6 Musketeer companies. This regiment was, however, made up of only one contingent. There were 2 3-pounder guns attached to each battalion, under the command of a lieutenant, and served by 2 or 3 gunners assisted by infantry men. The Upper-Rhenish gunners wore dark green coats, waistcoats and breeches with yellow hat lace and buttons.

31. The Empire: Officer, Franconian Infantry-Regiment Varell

This regiment was composed of 26 different contingents, one of which, from Ansbach, wore Prussian style mitre caps with a brass front, and a blue cloth back with a red lower part.

32. The Empire: Standard-bearer, Franconian Infantry-Regiment Ferentheil

This regiment consisted of 20 contingents including those from Nürnberg, Eichstätt, Hildburghausen, Bayreuth, and Würzburg. The standard-bearer shown here comes from this last contingent. The Würzburg Grenadiers had bearskin caps with brass plates, while those from Bayreuth had mitre caps with brass fronts and red cloth backs. The Franconian gunners wore blue coats with red collars, cuffs, lapels, and skirts; yellow hat lace and buttons; black cockades and red hat tassels; red waistcoats and breeches; and white waistbelts with a black pouch in front. Each battalion had a white *Leibfahne*, and 2 flags in the regimental colours bearing the Imperial eagle on one side and the initials F.K. (Franconian Kreis) on the other.

LEUTHEN 5 DECEMBER 1757

The victory of Rossbach had the effect of ruining the Franco-Reichs army, and henceforth Prince Ferdinand of Brunswick took the care of the French off Frederick's hands. However, the latter still had to deal with the Austrians and Russians who were pressing towards his capital. During his absence from Silesia the Austrians under Prince Charles of Lorraine and Daun had advanced and beaten a Prussian force at Breslau, and were on the way to making a complete conquest of the province. On 28 November Frederick linked up with Bevern's beaten troops, which were placed under Zieten's command, and on 4 December the whole force was concentrated at Neumarkt, near the Elbe, on the road from Liegnitz to Breslau. Frederick although outnumbered intended to risk all on one well-delivered blow. On the evening of 4 December, Frederick made his celebrated address to his generals in which he exhorted them to attack wherever the Austrians might be found. He threatened to dismount any cavalry regiment which failed to charge home and to strip away the colours, swords, and braid of any infantry regiment which so much as wavered.

The Austrian position lay across the high road to Breslau, resting on villages, swamps, and woody knolls; each flank was especially well defended.

At first light on 5 December the Prussian army moved off through a cold mist towards Leuthen. The Hussars of the advance guard had the first success of the day when they ousted the Austrian cavalry from the village of Borne. Almost immediately the main Austrian army emerged from its encampments, and formed a line, 4 miles long, between Nippern in the north and Sagschütz in the south. The village of Leuthen was located in the left centre of the line. The long lines of Austrians stood amazed as the little body of Prussians marched stolidly towards them. Laughingly, they called them the 'Potsdam Guard Mounting'.

Confronted by this solid wall of troops Frederick decided to attempt to turn the Austrian left flank, marching his army off at right angles along the Austrian front, partially screened by a line of low hills. At the same time he pretended to deploy his left wing thus fooling the Austrians into reinforcing their right flank.

Soon after noon, as the sun dispersed the mist, the Prussian van made a half left turn, which brought them into a position at right angles to the Austrian right flank. With great deliberation Frederick formed the army into two lines echeloned back to their left, pointing out the line of march in person to the Free Corporals of the leading regiments, whose colours would serve as marker points. A Free Corporal of the Infantry Regiment Meyerinck recalled having the entire Austrian army stretched out in front of him, and the entire Prussian army, drawn up in battle order, behind him. The Prussian army advanced 'with sounding music, as if on parade. Its order was as magnificent as at any review at Berlin, for it was marching under the eyes of its great monarch'.

The spearhead of the Prussian attack fell on the Württembergers and Bavarians, and in particular on the Württembergers regiments Spitznatz, Prinz Louis, Trüchfetz, and Leib. On the right the advance guard under Bevern, supported by 53 squadrons under Zieten repulsed a determined Austrian counter-attack. Artillery support was provided by the heavy guns which advanced rapidly from one position to another.

The Austrians fell back on Leuthen which became the central pivot for a re-alignment north-south facing the Prussian attack. In fierce hand-to-hand fighting in the churchyard the 3rd Battalion of the *Garde* turned out the Regiment of 'Red'-Würzburg. About 5 o'clock as the sun was beginning to set, the Austrian General Lucchese assembled 70 squadrons for a final attack on the Prussian left, but before he could set them into motion, they were themselves taken in the right flank by 35 squadrons of Prussian cuirassiers and 10 of the Bayreuth Dragoons. Thirty more squadrons from the Prussian reserve joined in, and the whole mass bore down on the Austrian infantry and broke them. Frederick pursued with three Grenadier battalions and the Seydlitz Cuirassiers, as far as Lissa. Prince Charles of Lorraine retired across the border leaving 17,000 troops stranded at Breslau, which surrendered on 20 December.

Frederick's 'oblique attack' was used at Leuthen to its finest effect, shattering not a mob of *Reichs* recruits but the newly reformed and equipped Austrian army. The fire power of the infantry was greatly increased by taking their ammunition waggons into action with them; and the artillery performed near miracles manhandling their 24-pounder *brummer* across country in search of new positions. Out of 60,000 the Austrians lost 21,000 including 12,000 prisoners and many guns. The Prussians lost some 6,400 out of 35,000. A battle map is given at the end of the book.

PLATE DESCRIPTIONS

33. Prussia: Private, Dragoon-Regiment No. 5. (Bayreuth)

The Bayreuth Dragoons, 10 squadrons strong, was probably the most celebrated and successful regiment in the Prussian army, particularly distinguished for the charge which decided the battle of Hohenfriedburg, in 1745. It also distinguished itself at Leuthen and Torgau.

34. Prussia: Officer, Dragoon-Regiment No. 1. (Normann)*

This regiment was 'remarkable for its bravery in all the Prussian campaigns'. Dragoon officers wore similar uniforms to their men, but of better quality. They had black cockades in their hats, and wore the Prussian officers' sash. The button-holes of their coats were frequently embroidered with *rococo* loops, as here.

35. Prussia: Drummer, Dragoon-Regiment No. 4. (Czettritz)**

Raised in 1741 from 5 squadrons of the 3rd Dragoons, this regiment distinguished itself at Rossbach and Zorndorf. Drummers of Dragoon regiments had lace of a special regimental pattern on their coats, 'swallow's nests', and drum carriages; and they carried infantry pattern side-drums.

36. Prussia: Drummer, Infantry-Regiment No. 10. (Pannwitz)**

A celebrated Westphalian regiment, which fought at Breslau, Leuthen, and Landeshut, where the 1st Battalion was captured. The Grenadiers served with those of Infantry-Regiment No. 9. Each infantry battalion had an establishment of 15 drummers and fifers who were responsible for playing the various field and camp calls. They were not used much on the march as the Prussians were proud of their ability to keep step without the assistance of a beat. The coats, 'swallow's nests' and drum carriages of the drummers and fifers were laced with special regimental lace.

37. Prussia: Standard-bearer, Infantry-Regiment No. 13. (Itzenplitz)**

This regiment, known under the regime of its fiery *chef*, General Itzenplitz, as the '*Donner und Blitzen*' regiment, fought at Lobositz, Prague, Rossbach, Leuthen and Hochkirch, where it suffered heavy casualties. Each infantry regiment had a Colonel's Colour, or *Leibfahne*, usually white and another colour, carried by the 1st Battalion, and a number of company colours, which were of the same design as the *Leibfahne* but of reversed colours. Apart from the 1st Battalion of the *Garde*, which had embroidered colours, all the others were painted. The colours were carried for the nominal Ensign by senior, or better-connected aspirant officers who were taken on in the rank of *Freikorporal*. These were expected to have full mastery of drill and regimental routine, and while having the status of NCOs were expected to cultivate the society of the officers.

38. Prussia: Oboist, Infantry-Regiment No. 15, Regiment-Garde, III Battalion

The 2nd and 3rd Battalions of the *Garde* fought at Rossbach, Leuthen (where the 3rd Battalion drove the Austrians out of the churchyard), Hochkirch, and Leignitz. The Grenadiers served with those of Infantry-Regiment No. 18. The 1st Battalion, known as the *Bataillon-Garde* – or *Leib-Garde-Bataillon* – wore a different and more elaborate uniform and was extremely efficient on parade in Potsdam although its fighting record was also good. It was recruited 'at large', mainly of foreigners. The 'Music' of the infantry consisted of the *Harmonie-Musik* combination of 2 oboes, 2 clarinets, 2 horns, and 2 bassoons. As yet the influence of percussive 'Janissary' music had not reached Prussia to change their delicate sound to the more familiar 'brass band sound'. In action the woodwind was posted behind the company colours which were massed between the 4th and 5th Platoons of the Battalion.

39. Prussia: Bombardier, Field Artillery Regiment

Frederick the Great inherited 1 battalion of artillery-men, 6 companies strong, which by the time of the Seven Years War had been increased to 2 battalions totalling 72 officers and 2,402 men. Bombardiers, of which there were 60 to 90 in each battalion, were responsible for serving howitzers and mortars. In 1742 2 light 3-pounder guns were allocated to each regiment, each served by 8 men, half from the artillery, and half from the infantry. On the shoulder belt which supported a circular wooden powder flask, the artilleryman carried 2 metal prickers for cleaning the vent of the gun.

40. Prussia: Jäger, Free Battalion No. 1. (Le Noble)

Raised in 1756 the Free Battalion Le Noble was captured at Landeshut, but was re-raised and a second battalion was added. At the Battle of Freiburg, in which the Free Battalions took a notable part, it appeared as a

Free Regiment. The *Jäger* here carries a short rifle and wears a leather cap or *Kaskett* similar in style to those adopted by the Austrian infantry after the Seven Years War.

41. Prussia: Jager, Jager-Corps zu Fuss

Formed in 1740 from 60 *Guiden* whose task was to protect reconnaissance parties, the foot jägers were increased to 2 companies of 100 men each. They distinguished themselves at Leuthen but in October 1760, having just been augmented to a complete battalion of 800 men, they were caught out in the open by a force of Cossacks, near Spandau, and virtually wiped out. The brown leather *Jäger* equipment consisted of a waistbelt with a small pouch in front. The men were armed with a sabre, a bayonet, and a *Büchse*, or hunting rifle. Officers wore gold aiguillettes and the silver Prussian sash; NCOs wore yellow and gold shoulder cords.

42. Austria: General Baron v. Loudon

Gideon Ernest Baron v. Loudon was born in Livonia in 1717, of Scottish descent. He served first in the Russian army but later joined the Austrian service. During the Seven Years War he was one of the liveliest of the Austrian commanders. After the war he was made Field-Marshal and commanded during the War of the Bavarian succession and the Turkish War (1788–9), during which he gained several victories and captured Belgrade. He died in 1790. Until 1751 Austrian General Officers had no regulation uniform but in that year a new uniform with carefully defined rank markings was introduced.

In this portrait General Loudon wears the star and ribbon of the Order of Maria Theresa, which was created in May 1757, a few weeks before the Austrian victory at Kolin, whose date (18 June) was afterwards adopted as the foundation date of the Order. It was awarded for military merit only.

Feldmarschall (F.M.) (Field-Marshal)	Whorls of gold lace down front of coat and around cuffs.
Feldzeugmeister (F.Z.M.) (General)	Three rows of gold lace around cuffs; one wide and one narrow on front of coat.
Feldmarschall-Lieutenant (F.M.L.) (Lieutenant-General)	Two rows on cuffs; one wide on front of coat.
General-Feldwachtmeister (G.-F.W.M.) (Major-General)	One row on cuffs; one narrow on front of coat.

43. Austria: Field-Marshal Prince Charles of Lorraine

Prince Charles of Lorraine, the brother of the Emperor Francis Stephen, was one of the leading Austrian commanders during the War of the Austrian Succession and the Seven Years War. In this portrait he wears the regulation uniform for Austrian field-marshals, with the Order of the Golden Fleece at his neck and the Star and Ribbon of the Order of Maria Theresa.

44. Austria: Engineer Officer

The Austrian army had a small body of 98 Military Engineers, dressed as shown here. In addition there was a pioneer corps and a sapper corps. The former formed in 1758, and consisting of 4 companies, fought at Hochkirch, Landeshut, before Glatz, and at the storming of Schweidnitz. Their uniform consisted of a black leather *Kaskett*, blue-grey single breasted coat and waistcoat, the former with green collar, cuffs, and turnbacks, white breeches and black gaiters. The Sapper Corps was raised in Dresden, in 1760, by the French General Jean-Baptiste Gribeauval. It consisted of 3 companies and served in the storming of Glatz and the various sieges of Schweidnitz. Their uniform was similar to that of the Pioneers except that they had red facings and blue-grey breeches. In addition their Kasketts had a yellow border, a yellow cipher of Maria Theresa on the front, and a red tuft at the side.

45. Austria: Trumpeter, Cuirassier-Regiment O'Donell

Raised in 1721 this regiment fought at Lobositz, Breslau, Leuthen, Hochkirch, Meissen, Maxen, Landeshut, Leignitz and Torgau. Cuirassier regiments had kettledrummers as well as trumpeters.

46. Austria: Officer, Cuirassier-Regiment Anhalt-Zerbst

Raised in 1682 this regiment fought at Prague, Breslau, Leuthen, Hochkirch, where it distinguished itself, Maxen and Torgau. Austrian officers generally wore similar uniforms to their men but of better quality, with the addition of black and gold hat tassels, sash and sword-knot. The ranks of regimental officers were as follows: *Obrist* (Colonel), *Hauptmann* (Infantry Captain), *Rittmeitser* (Cavalry Captain), *Lieutenant*, *Fähnrich* (Ensign), and *Kornett* (Cornet).

47. Austria: Private, Cuirassier-Regiment Serbelloni

Raised in 1672, this regiment fought at Lobositz and Kolin with distinction, and at Schweidnitz, Hochkirch, Meissen, Maxen, and Torgau.

48. Austria: Private, Dragoon-Regiment Württemberg

Raised in 1688, this regiment fought at Kolin with distinction, at Breslau, Leuthen, and Olmütz. Part of it was present on 30 June 1758 when a force under Loudon virtually destroyed a Prussian supply column of 4,000 waggons near Domstadl. It was also present at Kunersdorf, Landeshut, and Leignitz. There were 14 Dragoon regiments in the Austrian army, organised like the Cuirassiers. They retained, however, several infantry features. Thus their captains were called *Hauptmann*, they had bayonets on their carbines, used infantry side drums, and carried infantry colours. During the period of army reforms Maria Theresa wanted to clothe the Dragoons in 'Imperial' white with blue facings to distinguish them from the Cuirassiers, but the opposition from regimental proprietors was too strong and the scheme was abandoned.

49. Austria: Officer, Dragoon-Regiment Zweibrücken

Raised in 1701, this regiment fought at Reichenberg, Jungbunzlau, Prague, Kolin, Breslau, Leuthen, Olmutz, Domstadl, and Hochkirch. This officer wears the normal uniform of his regiment with the addition of the Austrian black and gold sash.

50. Austria: Drummer, Dragoon-Regiment Löwenstein

Raised in 1758, this regiment fought at Olmütz, Holitz, Stolpen, Hochkirch, Kunersdorf, Landeshut, and Freiberg.

51. Austria: Sergeant, German Infantry-Regiment Ligne

Raised in 1725, this regiment fought at Hastenbeck (one battalion only), Kolin, Moys, Breslau, and Leuthen, where it was one of the reserve regiments hurled into Leuthen village at the height of the battle. It was also present at Pirna, Sonnenstein, Hochkirch, Maxen, Berlin, Burkersdorf, and Schweidnitz. At the outbreak of the Seven Years War Austria possessed 56 regiments of infantry, of which 39 were 'German', 11 Hungarian, 4 Netherlandish, and 2 Italian. Each regiment consisted of 4 battalions, but 3 only served in the field divided into 2 'field-battalions' of 6 companies each. In action the battalion was divided into 'divisions' of 4 platoons each. The strength of the company was 3 officers, 6 NCOs, 3 drummers, one fifer, and 130 privates. The infantry wore a basically white uniform with different regimental facing colours, but there were many minor variations in such things as hat lace, pom-poms and tassels, shoulder-straps, and the small cloth loops which joined the skirt turnbacks. The sergeant carries a halberd cane and sword. Red stocks were worn by the German regiments, although there were exceptions, and black by the Hungarian regiments.

52. Austria: Grenadier Officer, German Infantry-Regiment Arberg

Raised in 1742, this regiment fought at Prague, Moys, Breslau, Leuthen, where it was part of the reserve, Hastenbeck (3rd Battalion only), Hochkirch, Landeshut, Leignitz, and Schweidnitz. The two Grenadier companies of each regiment were usually taken away in wartime and massed with others to form a Grenadier corps. The Grenadiers, who received higher pay, wore fur caps, carried matchcases on their cartouche belts, and were armed with a sabre in addition to their muskets and bayonets. The officer shown here has a fusil and bayonet instead of an espontoon, and he wears a waist-pouch in addition to his officer's sash.

53. Austria: Musketeer, German Infantry-Regiment Merci

Raised in 1684, this regiment fought at Reichenberg, Prague, Kolin, at Breslau with distinction, Leuthen, where it was part of the reserve, Moys, Neisse, Torgau, Schweidnitz and Burkersdorf. The Musketeers, or Fusiliers, were armed with the 1754 pattern musket and bayonet.

54. Austria: Musketeer, German Infantry-Regiment Charles of Lorraine

Raised in 1715, this regiment fought at Prague, Schweidnitz, Leuthen, where it distinguished itself at Windmill Hill, Hochkirch, Breslau, Dresden, Torgau, and Schweidnitz. The private shown here carries a knapsack and waterbottle on separate straps over his right shoulder. The scalloped hat lace and the oblong skirt ornament are typical regimental distinctions.

55. Austria: Ensign, German Infantry-Regiment Hoch u. Deutschmeister

Raised in 1695 from the remnants of the household troops of the Grand Master of the Teutonic Order, Franz Ludwig of Pfalz-Neuberg, its colonels continued to be the successive Grand Masters of the Order. It fought at Kolin with distinction, Schweidnitz, Breslau, Leuthen, Leignitz, and Landeshut. Each regiment had a Colonel's Colour, or *Leibfahne*, and 3 battalion colours, known as *Ordinaires*. They were rectangular in form 1.8 m. × 1.4 m. with a metal spearhead. The *Leibfahne* was white with a representation of the Virgin Mary on the obverse, and the Austrian double-headed eagle on the reverse; it was usually embroidered. The battalion colours were of yellow silk, bordered on three sides with black, yellow, white and red triangles (the colours of the German Empire). On both sides was painted the Austrian double-headed eagle.

56. Austria: Drummer, German Infantry-Regiment Luzan

Raised in 1721 this regiment fought at Schweidnitz, Breslau, Leuthen, Hochkirch, Dresden and Strehlen. The 3 drummers and one fifer from each company provided the *Feldmusik* of the regiment.

57. Austria: Grenadier, Hungarian Infantry-Regiment Josef Esterhazy

When Maria Theresa came to the throne there were only 3 Hungarian infantry regiments, but 8 more were soon raised, although with considerable difficulty for while the Hungarians had no objection to being Hussars of freebooters they were not much inclined to serve as regular infantry. This regiment, raised in

1741, fought at Lobositz, Pirna, Prague, Breslau, Leuthen, at Hochkirch with distinction, Maxen, Meissen, Landeshut, Leignitz, and with distinction at Schweidnitz. In the Hungarian regiments the coats were white, like the Germans, but without lapels and with coloured loops. The waistcoat was corded in the Hungarian or 'Hussar' manner. The waistcoat and tight pantaloons were usually of the facing colour, except in regiments with yellow facings, when they were bright blue. Over the waistcoat was worn the hussar 'barrel' sash, and some regiments had sabretaches. All the Hungarian infantry carried a curved sword.

58. Austria: Corporal, Hungarian Infantry-Regiment Gyulay

Raised in 1702, this regiment fought at Prague, Moys (grenadiers only), Berlin (1 battalion), Meissen, Maxen, Kunersdorf, Torgau, Schweidnitz, and Freiberg. Corporals carried canes and halberds with heads which differed in shape from those of the sergeants (see fig. 51).

59. Austria: Grenadier, Hungarian Infantry-Regiment Haller

This Grenadier, after a painting by Morier, carries a match case on his cartouche belt, and wears *charivari*, or thigh length over-breeches, typical of Hungarian dress.

60. Austria: Gunner, German Field Artillery Corps

Between the wars the Austrians, unknown to Prussia, succeeded in building up a powerful reserve of fine new artillery pieces, and in training up a corps of highly professional gunners. The artillery personnel came from the German Field Artillery Corps, and the Netherlands Field Artillery Corps. The former consisted of 3 brigades of 8 companies each, increased to 10 by 1760. Each company consisted of a captain and 70 qualified gunners (*Büchsenmeister*) a number which had increased to 130 by 1760. The Netherlands Artillery, 600 men strong, wore similar uniforms to the German Artillery, but with red lapels and turnbacks. Both units played an important part in the various engagements throughout the war.

61. Württemberg: Grenadier, Infantry Regiment Leib

One of the regiments upon which the brunt of the Prussian initial attack at Leuthen fell. There were 5 Württemberg infantry regiments, each of 2 battalions present at the battle.

62. Bavaria: Musketeer, Infantry Regiment Leib

There were 5 Bavarian regiments present at Leuthen, each of 2 battalions.

MINDEN 1 AUGUST 1759

In March 1757 a combined Franco-Austro-Imperial army of 100,000 men, under the Prince de Soubise, crossed the Rhine threatening Hanover. The so-called 'Army of Observation', consisting of 47,000 Hanoverians, Hessians, and Brunswickers awaited its approach without a leader, and at the insistence of Frederick the Great and George II, the reluctant Duke of Cumberland was persuaded to take command. He arrived at Stade on 14 April, when the campaign had already opened, and three months later, on 26 July, was defeated at Hastenbeck. Forced to retreat on Stade on 5 September he concluded the Convention of Klosterseven, whereby he agreed to evacuate the country. The news of this came as a blow to Frederick who appealed to George II not to abandon him to his enemies. At this time the Russians were ravaging East Prussia, Silesia was overrun by the Austrians, and Soubise's army in Westphalia was about to join up with that commanded by the Duc de Richelieu, which had just defeated Cumberland.

The British government eventually informed Frederick that the Convention would not be ratified and requested the services of Prince Ferdinand of Brunswick as commander-in-chief of the Army of Observation. After a certain amount of negotiation over the terms of his command Ferdinand took over his disorganized and demoralized army on 23 November, 1757, and promptly set about putting it into some sort of order. As this work went on throughout the winter Ferdinand repeatedly requested that a contingent from Britain be sent to Germany to protect his right flank. Early in 1758 a small naval force entered the Weser, made contact with Ferdinand, and captured the Port of Emden. The British 51st Regiment of Foot was sent out soon after to garrison the port, the vanguard, as it turned out, of a small British expeditionary force of 17 battalions and 13 regiments of horse, under the command of the Duke of Marlborough, which joined Ferdinand at Coesfeld on 22 August 1758. There had been much sickness among the newly arrived troops and on 20 October the Duke of Marlborough died, apparently of dysentery, and was succeeded by Lord George Sackville. The campaign of 1758 came to an end in November, with Ferdinand having successfully defended Frederick's right flank against at least three opposing armies. He had even managed to inflict a severe defeat on the French at Crefeld, on 23 June. The campaign of 1759 started early when, in January, Soubise, with 31,000 men, seized the neutral and free city of Frankfurt, by guile, thus gaining mastery of the Rhine and Main. There followed months of manoeuvring as Ferdinand gradually fell back before superior forces, yielding up the whole of Westphalia and Hesse. On 10 July the town of Minden, on the Weser, fell to de Broglie's army, with all its stores and artillery, while Ferdinand was miles away near Osnabrück. Contades joined Broglie with his army and the way to Hanover and Brunswick was open.

Ferdinand hurried to block this threat and found Contades in a very strong position north of Minden, with his right on the Weser, his left on a steep hill, and to his front an extensive marsh. De Broglie was posted across the river to protect his right flank. In order to tempt him out Ferdinand formed his army in a long curve to the north-west of Minden, in such a way that Wangenheim's Corps, on his left wing, appeared to have been left in mid-air. At the same time the 'Erbprinz' (Ferdinand's nephew, the Hereditary Prince of Brunswick) was detached to threaten the French communications. The ruse worked and on the night of 31 July de Broglie broke camp, crossed the Bastau brook, and marched to attack Wangenheim, while Contades advanced the main army over the Bastau marsh and deployed on the extensive heath to his front. The allied outposts were not very alert and Ferdinand lost several hours before a deserter informed him of the French move. Luckily for him, however, de Broglie, who was formed up by 4 o'clock in the morning, delayed his attack and so lost the advantage of surprise.

At about 5 o'clock the allies advanced in eight parallel columns towards the French main army which was drawn up in an acre with the infantry and artillery on each flank, and the cavalry in the centre, where the ground was hardest. Through some misunderstanding of its orders the two British brigades, in the third column from the right, set off before the battle line had been properly formed. With three Hanoverian battalions on their left, the six British battalions stepped out towards the astonished French with drums beating and colours flying. They were temporarily halted by frantic staff officers but soon set off again bearing down on the French cavalry, 7,000 men in 63 squadrons, and for a space of some 150 yards passed through the cross-fire of 66 guns. They halted just short of the French cavalry and met their repeated counter-attacks with a hail of bullets that swept them off the field. 'Never', wrote one eye-witness, 'have so many boots and saddles been seen upon a battlefield as in front of the English Infantry and the Hanoverian Guards'. Contades tried various means to shift them. A force of 32 guns and four infantry brigades was sent round their right flank but was seen off. An attack by seven of Prince Xavier's Saxon battalions almost succeeded in a similar attempt, marching twice over the unhorsed General Kingsley before they were repulsed and he was able to escape. An attempt by de Broglie to draw Ferdinand's attention away to his left flank was crushed by a charge of the German cavalry under the Duke of Holstein-Gottorp, and Contades ordered a withdrawal.

All this time Ferdinand had been sending orders to Lord George Sackville who was at the head of the British cavalry, to charge the enemy to his front, but for some reason, attributed by one historian to 'spleen and bad temper, if not actual cowardice' he hesitated, only finally moving off, after repeated orders, and when it was too late. By 10 o'clock the French, in disorder, were pouring back across the Bastau and through Minden. It was due largely to Ferdinand and his Anglo-Germans that Frederick recovered the Rhenish and Westphalian territories he had left without any protection, and that Hanover and Brunswick were preserved. The allies lost some 2,800 out of 42,500, half of them from the British infantry and artillery. The French lost 7,000 out of 54,000, as well as most of their baggage, 17 standards and colours, and 43 cannon.

PLATE DESCRIPTIONS

63. Hessen-Cassel: Private, Cavalry-Regiment Leib

There were 4 regiments of Hessian cavalry at Minden, each 2 squadrons strong. They formed part of the eighth or extreme left column of the Allied army, which consisted of German cavalry under the command of the Duke of Holstein-Gottorp. Unlike the British cavalry they showed no reluctance in charging the right of the French line towards the end of the battle. The Hessian regiments all wore white uniforms with different coloured facings. The details of this figure are taken from a painting of a private by Morier.

64. Hanover: Private, Horse Grenadiers

The Horse Grenadiers consisted of 2 companies totalling 8 officers, 14 NCOs, 4 drummers and 150 grenadiers. Together with the *Gardes du Corps* they were the only Hanoverian cavalry to wear red. At Minden, together with the *Gardes du Corps*, the cavalry regiments of Breitenbach and Bremer (see fig. 162), and the Beltheim Dragoons, they formed part of the first or extreme right column of the Allied army consisting of cavalry under the command of Lord George Sackville. The Hanoverian *Gardes du Corps* consisted, in 1756, of 3 companies of 9 officers, 12 NCOs, 4 trumpeters and 156 privates.

65. Great Britain: Grenadier, 12th Regiment of Foot (Napier)

This and the following five figures re present the six 'Minden' regiments, which were formed into 2 brigades, the first, Waldegrave's, containing the 12th, 23rd, and 37th Regiments, and the second, Kingsley's, the 20th, 25th and 51st. In June 1759 Prince Ferdinand formed their 6 Grenadier companies into a battalion commanded by Major Maxwell of the 20th Foot. During the Battle of Minden, Maxwell's Grenadiers were part of Wangenheim's Corps. Raised in 1685, this regiment was numbered as the 12th when numbers were first allocated in 1751, although the use of the Colonel's name lingered on until at least the end of the war. It took part in most of the major actions of the campaign in Germany, including Warburg, Klosterkampen, Vellinghausen, where they received the surrender of the French Infantry Regiment Rougé, complete with its guns and colours, Wilhelmstahl, Lutterberg, and Cassel. By Royal Warrant of 15 October 1755, each battalion was to have a Grenadier company consisting of 4 sergeants, 4 corporals, 2 drummers, 2 fifers, and 100 grenadiers. The Grenadiers were distinguished by their tall mitre caps, matchcases on their cartouche belts, and shoulder-wings, ordered for all regiments in 1752 with the exception, according to Morier, of the 12th Regiment. An extra pouch was sometimes worn on the waistbelt. The hair was worn plaited and tucked up underneath the cap.

66. Great Britain: Officer, 23rd Royal Welsh Fusiliers

Raised in 1689, this regiment was selected, during the War of the Spanish Succession, together with an English and a Scottish regiment, to become 'Fusiliers' for the purpose of guarding the artillery train, a practice copied from the French. At the outbreak of the Seven Years War it was one of the 4 regiments which unsuccessfully defended Minorca against the Duc de Richelieu. Afterwards it took part in the raids on Rochefort and St Malo and in the campaign in Germany being present at Minden, Warburg, Klosterkampen, Vellinghausen, and Wilhelmstahl. The Fusiliers, with the exception of the officers who seem to have worn hats, all wore mitre caps. Originally they were ordered to be lower than those of the Grenadiers, but this distinction seems to have been soon lost. The officers seem to have carried espontoons at least at the beginning of the Seven Years War (but were soon discarded as 'burdensome' in favour of the fusil and bayonet). The signs of officer rank were the shoulder cords on the right shoulder, intended to hold the sash in place, the gorget, the crimson sash, and the crimson and gold sword-knot. In 1747 officers were ordered to wear 'queued wigs or their hair done in the same manner'. Pigtails seem occasionally to have been worn by the officers but not by the men, who wore their hair tightly brushed up and tucked under the hat. The battalion, less the Grenadier company which was usually detached on service, was formed into 4 'grand divisions', of 2 companies each, sub-divided for tactical purposes into 3 equal platoons. Each platoon was commanded by a captain, lieutenant, and ensign with a sergeant to each. The remaining officers and NCOs were posted at the rear, 'to complete the files, to keep the men in their duty, and to supply the place of the officers and serjeants that may be killed or dangerously wounded'.

67. Great Britain: Private, 37th Regiment of Foot (Stewart)

Raised in 1702, this regiment took part in the expedition to Rochefort, Minden, Warburg, Klosterkampen (Grenadiers only), Vellinghausen, and Wilhelmstahl (Grenadiers). This figure is based on a sketch of Corporal Jones of the 13th Foot, drawn by Lieutenant William Baillie in 1753. The marching order of a British infantryman, in 1762, quoted by Fortescue, consisted of his coat weighing 11 lbs, his knapsack with contents, *viz.*: 2 shirts, 2 stocks, 2 pair stockings, 1 pair summer breeches, 1 pair shoes, brushes and blackball, totalling 7 lbs 10 oz. Other items and 6 days provisions amounted to 39 lbs 7 oz, the whole totalling 63 lbs 3 oz.

68. Great Britain: Fifer, 25th Regiment of Foot (Home)

Raised 1688, this regiment took part in the expedition to Rochefort and St Malo, Minden, Warburg, Zenenberg, Klosterkampen, Vellinghausen and Wilhelmstahl. The establishment for a battalion, of 1755, allowed for 2 drummers per company, and 2 drummers and 2 fifers for the Grenadiers. In action the drummers were 'to stay with their respective companies, to assist the wounded men'. James in his *Military Dictionary* lists the different drum beats as:

The *assembly*, the *general*, 'to give notice that the troops are to march', the *march*, 'to command them to move with the left foot first', the *tattoo*, or *tapto*, 'no more drinks, all to retire to their quarters', *reveillé*, *to arms*, *the retreat*, *the alarm*, *the parley*, and *the chamade*. Drummer and fifers, except for the Foot Guards and Royal Regiments, were to wear coats of 'the colour of the facing of their Regiments, lined, faced, and lapelled on the breast with red, and laced in such manner as the Colonel shall think fit'. The drums were to have the front of the shell painted in the facing colour, 'with the *King's Cipher* and *Crown*, and the *Number* of the Regiment under it'. The Fifer here has a coat with hanging sleeves, a traditional feature of musicians' livery, and a brass fife case suspended from a cord over his left shoulder.

69. Great Britain: Officer, 51st Regiment of Foot (Brudenell)

This regiment was one of the 10, numbered 52–61, ordered to be raised at the end of 1755, and was at first numbered 53rd. With the disbandment of the remnants of Shirley's (50th) and Pepperells' (51st), they were renumbered 51st early in 1757. The regiment was present at Rochefort, and was sent out to garrison Emden in 1758. It was present at Minden, Corbach, Warburg, Klosterkampen, Vellinghausen and Wilhelmstahl. The officer shown here is based on a portrait of Captain William Baillie who commanded the Grenadier company of the 51st during the campaign in Germany.

70. Great Britain: Corporal, 20th Regiment of Foot (Kingsley)

Raised in 1688, this regiment took part in the abortive raid on Rochefort, in September 1757, and in that on St Malo in May of the following year. At Minden the regiment is traditionally supposed to have been posted in or near some rose gardens, from which the men picked flowers to deck their hats. The regiment was also present at Warburg, Zierenberg, Klosterkampen, Vellinghausen and Wilhelmstahl. Shortly before the war the regiment was commanded by James Wolfe who left some interesting orders related to the training of the regiment. In these he mentioned a 'piquet' company which took post on the left flank of the battalion. As Wolfe had spent some time in France he may have copied this custom from them; it was obviously the forerunner of the light company. The figure shown here has a corporal's white tape knot on his right shoulder, an additional pouch on his waistbelt and heart-shaped skirt ornaments.

71. Hanover: Grenadier, Infantry-Regiment Hardenberg

This regiment was numbered 6A in the Hanoverian system. At the outbreak of the Seven Years War the Hanoverian infantry consisted of a Guards regiment of 2 battalions and 23 line regiments each of 1 battalion, divided into a Grenadier company of 1 officer, 4 NCOs, 2 drummers, 2 fifers, 3 pioneers, 7 *Gefreiters*, and 41 grenadiers, and 7 Musketeer companies of 3 officers, 8 NCOs, 2 drummers, 14 *Gefreiters*, and 87 privates each. In 1758 this establishment was increased by 14 men per company. In 1759 the Grenadiers were formed into 3 Grenadier battalions. Each battalion had 2 colours, a white *Leibfähne*, and another which varied with each regiment. At the beginning of the war the coats were laced as shown here, after a painting by Morier. Although Hardenberg took part in the famous advance at Minden in fact it was the right hand regiment of the 4th Column under Major-General v. Scheele, which had become separated.

72. Hanover: Officer, Regiment Garde

This regiment had 2 battalions at Minden which took part in the advance with the remainder of Spörckern's column. Based on a portrait, this officer has the usual marks of his rank, espontoon, gorget, and sash, in this case yellow for Hanover. The rank and file had their coats laced in yellow and their Grenadier caps had blue fronts, with a silver Garter Star surmounted by a gold crown, with a small red flap bearing the white Horse of Hanover.

73. Hanover: Sergeant, Infantry Regiment Spörckern

Numbered 2A, this regiment was part of Wangenheim's Corps at Minden. This figure has the usual marks of a sergeant, halberd, gloves, and yellow Hanoverian sword-knot.

74. Brunswick: Grenadier, Infantry-Regiment Imhoff

This regiment consisted of 2 battalions and was in Lieutenant-General v. Imhoff's column, the seventh from the right at Minden. The Brunswick Grenadier battalion was also in Wangenheim's corps. This figure which is based on a painting by Morier shows the usual Grenadier features. The other Brunswick line regiments, Behr and Zastrow had, respectively, red facings with white buttons, and yellow facings with white buttons. The Musketeers had hats with lace of the button colour. The officers wore silver and yellow Brunswick sashes, and their uniforms in general followed the Prussian style.

75. Hesse-Cassel: Officer, Infantry-Regiment Garde

Raised in 1631, this regiment, of 1 battalion, was in the 6th Column at Minden, under Lieutenant-General v. Wutginau. The figure shown here is based on a painting of the Landgraf's aide-de-camp, who appears in the hunting scenes painted by Tischbein c. 1760 and now in the *Reihersaal* at Schloss Fasanerie. The uniform is very Prussian in style with a single-breasted coat and embroidered button-holes similar to those worn by the Prussian, and Brunswick Foot Guards. The Hessian sash was silver and red. The rank and file of the *Garde* wore silver-fronted Grenadier caps, white shoulder cords, and white lace with a red stripe.

76. Schaumberg-Lippe: Musketeer, Infantry-Regiment

The contingent from Schaumberg-Lippe was raised from men living on the territory of the Graf (see fig. 78), which lay in a bend of the River Weser, south-east of Minden. It was small but well-trained under the personal supervision of the Graf, and consisted of an infantry battalion used principally to escort the artillery, an artillery detachment with 28 guns, and a small regiment of light horse, the Bückeburg Carbiniers.

77. Great Britain: Gunner, Royal Regiment of Artillery

The Royal Artillery dates from 26 May 1716, when two companies were permanently established at Woolwich. By 1754 there were 5 companies serving abroad, 5 at Woolwich with the Company of Gentlemen Cadets at the Royal Military Academy at Woolwich which had been founded in 1741. The Royal Artillery, which was part of Spörckern's Column, rendered valuable service at Minden, and elsewhere in Germany, Canada, Belle-Isle, and Havana. Certain companies were also employed in the East-Indies before 1763. The figure shown here has a carbine and bayonet and a powder-horn worn on a shoulder-belt. Sergeants had gold lace and shoulder-cords and corporals yellow worsted. Corporals and bombardiers carried carbines instead of halberds. An officer of the Royal Artillery is shown at fig. 138.

78. Graf v. Schaumberg-Lippe-Bückeburg

Friedrich Ernst Wilhelm Graf v. Schaumberg-Lippe, the 'greatest artilleryman of his age', was born in London in 1724, reputedly an illegitimate son of George I. After short service in the British and Austrian armies he succeeded, in 1748, as reigning Graf. After some years studying the military profession in Prussia, he returned to Bückeburg and introduced compulsory military service. In 1756 he was appointed *General Zeugmeister* in the Hanoverian army and was present at Hastenbeck, Lutterberg, and Minden. He directed the siege operations at Wesel in 1760 and Cassel in 1761, and was present at Vellinghausen. In 1762 he became a Field-Marshal in the Portuguese service, returning to Bückeburg in 1763, where he founded a military academy, at which Scharnhorst was a pupil. He died in 1777. In his portrait by Sir Joshua Reynolds he wears a very simple uniform with the ribbon and badge of the Prussian Order of the Black Eagle.

79. Hanover: Gunner, Hanoverian Artillery

At the beginning of the Seven Years War Hanover had 6 companies, each of 50 gunners, which were increased in 1756 by 2 more. In 1758 a further 2 were added. At Minden the artillery formed the 2nd Column, under Major Haase, with 30 guns and howitzers. Each Hanoverian infantry battalion had 2 3-pounder guns in the charge of 1 officer and 15 artillerymen. In addition there were 6 ammunition waggons and other vehicles, 24 horses, and 8 drivers. The Hanoverian guns and waggons were painted red. The gunner here carries a powder flask on a belt over his left shoulder, and a linstock, with slow-match twisted round the pole, used for firing the guns.

80. France: Gensd'armes Anglais du Roi

The *Gendarmerie de France* were part of the French Household troops and were of great antiquity. In 1665 they were put on a regular footing, being formed into 17 companies each of 250–300 gentlemen volunteers. At the time of the Seven Years War the *Gendarmerie* consisted of 10 companies, 4 of which, including the 1st Scottish and 2nd English, belonged to the king, and the rest to the queen and various princes. Eight companies were at Minden acting as a reserve to the cavalry in the centre. 'I only know about the gendarmerie', Prince Ferdinand is reputed to have said about them after Minden, 'because of their courage. At the Battle of Minden I saw 30 of their officers lying on the field of battle'. Each company was commanded by a captain-lieutenant, a sub-lieutenant, and an ensign, all of whom ranked as colonels in the army, and a guidon or cornet, who ranked as a lieutenant-colonel. Of the warrant officers the *maréchal-de-logis* ranked as a captain, and the *brigadier* as a lieutenant. There was a servant to every 4 *gendarmes* who looked after them and the pack horse which carried their equipment on campaign. The various ranks were shown by the amount of silver lace on the edges and seams of the coat, and the different companies by the central colour of their carbine belts, and the insignia on their pistol holders and saddle cloths. They were armed with a carbine, 2 pistols and sword, and wore *calotties de fer* in their hats and breast-plates on service. The *Gendarmerie Anglais* were formed in 1667 by Sir George Hamilton of Abercorn from catholics in Charles II's Household Cavalry. About 1740 they took violet as their distinguishing colour and a double AA Cipher surmounted by a crown.

81. France: Private, Cavalry Regiment Colonel-Général

The senior of the 'General-Staff' regiments, and ranking No. 1 of the cavalry regiments, the Regiment Colonel-Général was formed in 1635. As senior regiment it was on the right of the front line of the cavalry at Minden. The private here wears a coat with long lapels down to the hem, although the 1761 manuscript shows the regulation coat with short lapels. While the regulations give white epaulettes, the Prince de Turenne seems to have succeeded in keeping his black and white check livery lace.

82. France: Private, Cavalry Regiment Cuirassiers du Roi

Raised in 1666 and numbered seventh in seniority, this regiment was part of the first line of de Broglie's corps at Minden. This figure, after a painting by Le Paon, shows the characteristic bearskin Grenadier cap, which the regiment seems to have adopted before 1750, and the single shoulder belt for carbine and pouch.

83. France: Officer, Cavalry-Regiment Royal-Cravattes

Taken into French service in 1667, and eighth in seniority, this regiment was on the right of the second line of cavalry at Minden.

84. France: Private, Cavalry Regiment Royal-Pologne

Raised in 1653 and numbered 13th, the regiment was in the front line of the cavalry at Minden. The only regiments which were allowed to wear bearskin caps, according to the regulations, were the German regiments of Royal-Allemand, Württemberg, and Nassau-Saarbrück, but like the Cuirassiers du Roi, the Royal-Pologne seem to have adopted them unofficially. The 1761 manuscript shows the Polish coat with lace brandenbourgs and pointed cuffs, and is the only authority to show the uniform of the regiment at this time.

85. France: Private, Schomberg Volunteers

The celebrated *Voluntaires de Saxe*, founded by Marshal Saxe in 1743 consisted of Uhlans, dressed in Tartar costume, and Dragoons, dressed in German fashion, but with the fantastic metal helmet, which seems to have been a particular invention of the Marshal and formed the pattern for the characteristic French heavy cavalry helmet worn later. After the death of the Marshal in 1752, the Uhlans were disbanded. The Dragoons were kept on, and at Minden were part of the rear line of de Broglie's corps. They were also present at Hastenbeck, Crefeld, Bergen, Corbach, Warburg, and Wilhelmstahl. After the war they became the 17th Regiment of Dragoons.

86. France: Private, German Hussar-Regiment Royal-Nassau

Hussars under Louis *X*V were without exception foreigners. After the Peace of Aix-La-Chapelle the numerous bodies of Hussars were reorganized, and an Order of 15 May 1752 regularized the uniform of the 7 remaining regiments. Shortage of recruits, desertion, and general misconduct reduced the Hussars to a weak state in the opening months of the Seven Years War and in 1756 the German Regiment of the Prince of Nassau, 4 squadrons of 150 men each, was taken into French service. At Minden it was part of de Broglie's corps. Its uniform was similar to that of the other French Hussar regiments and its arms consisted of a curved sabre, 2 pistols and a carbine.

87. France: Drummer, Infantry-Regiment Rouergue

Raised in 1667 and numbered 42nd, this regiment was in the front line of Contades' corps at Minden. Each of the 17 companies in the French infantry battalion had a drummer, but there was only 1 fifer to the battalion, and no woodwind players at all. Apart from some 25 regiments which wore the livery of their colonel (see fig 146), drummers wore blue coats with the red and white French Royal livery lace. The fronts of the drums were painted with a trophy which usually included the arms of France and an arrangement of the Colours of the regiment.

88. France: Grenadier Officer, Infantry-Regiment Auvergne

One of the so-called '*Petit Vieux Corps*', this regiment was raised in 1661 and numbered 9th. At Minden it was in the second line of Contades' corps. The figure here wears the uniform of the regiment with a fur cap, waist-pouch, fusil and bayonet.

89. France: Fusilier, Infantry-Regiment du Roi

Raised in 1665 and numbered 12th, this regiment was on the left of the line of Contades' corps at Minden. It was one of the few regiments which had lace brandenbourgs on the front of their coats.

90. France: Fusilier, German Infantry-Regiment Deux-Ponts

Raised for French service by the Prince Palatine of Deux-Ponts (or Zweibrücken) this regiment was numbered 92nd. At Minden it was in the third line of de Broglie's corps. In common with the other German regiments in French service it wore light blue coats.

91. France: Private, Grenadiers de France

After the Peace of Aix-la-Chapelle, when 53 French infantry battalions were disbanded, it was decided to retain their *compagnies d'élite* and to reorganise them as a corps called the *Grenadiers de France*. Composed of 4 brigades of 12 companies and numbered the 40th the Grenadiers de France and the Grenadiers Royaux had between them 8 battalions at Minden; 'as fierce and terrible-looking fellows as I ever saw', was one British eyewitnesses' description of them. At Wilhelmstahl 635 of their number were taken prisoner by the 5th Foot who are supposed to have taken their caps into wear thus introducing bearskin caps into the British army. A watercolour dated 1754 shows a different uniform without the white brandenbourgs which was possibly an undress coat worn for everyday duties and on campaign.

92. Saxony: Fusilier, Infantry Regiment Lubomirsky

After the disaster at Pirna in October 1756, a new Saxon army was raised to fight with the French, and by 1758 there were 9,000 Saxons, under Prince Xavier, Comte de Luscace, serving with Contades. At Minden the 15 Saxon battalions on the left of the second line mounted a strong attack against the right of the advancing British infantry regiments. The Saxon infantry all wore white uniforms, of Prussian style and cut, with different coloured facings, except for the Grenadiers of the *Garde* who wore red.

KUNERSDORF 12 AUGUST 1759

The victories at Rossbach and Leuthen enabled Frederick to recover those parts of Silesia which had been occupied by his enemies. In 1758 Olmütz was successfully besieged by the Prussians, after which Frederick moved east and defeated the Russians at Zorndorf, in August. He then hurried back to Saxony and manoeuvred to bring the cautious

Daun to action. The latter ultimately ventured to attack and surprised, and drove back, Frederick's army, at Hochkirch, in October, without important results. Throughout the rest of the year, and most of 1759, Frederick played a waiting game, at length venturing out to attack the Russians who were occupying a strong position on the far bank of the Oder, due east of Frankfurt.

The Russian army occupied three heights, separated by ravines, namely, from west to east, the Judenberg, the Laudongrund ravine, Grosser Spitzberg, the Kuhgrund ravine, Mühlberg, and the Beckergrund ravine. The position was 6,000 yards long, and the height reduced progressively from west to east. Near the village of Kunersdorf there were some ponds; and to the north, in front of the heights, the country was marshy with the Hühnerfluss stream flowing through it. The troops of Fermor and Villebois were on the Judenberg, Roumiantzov's on the Grosser Spitzberg, and Galitzine's on the Mühlberg. The Austrian infantry was posted at the foot of the north face of the Judenberg, near Rothvorwerk; and the Austrian cavalry nearby at the bridge over the Oder. The baggage was on the river bank guarded by a regiment of infantry. As at Zorndorf the ravines cut the Russian line into several parts. The Observation Corps, composed of the weakest units, was posted on the lowest height, the summit of which was commanded by the neighbouring hills.

Frederick was expected to attack from the north, but he made a long and difficult march through woods and marshes and across the Hühnerfluss, eventually appearing to the south; but having detected this movement Saltykov made a rapid change of front transferring part of his artillery to the southern points. Three grand batteries were erected on the Judenberg, a large and a small battery on the Spitzberg, and two more on the Mühlberg, all joined by trenches from which the infantry and artillery could fire. The troops of Villebois crossed the Spitzberg, the Austrian infantry climbed up the Judenberg, and the cavalry drew closer to the heights.

At dawn the Austrians set fire to Kunersdorf. At 9 o'clock the Prussian artillery opened fire from the heights of Trettin, occupied the Kleiner Spitzberg, and bombarded the Mühlberg from several sides, while the infantry formed four lines and the cavalry four groups. Saltykov passed the Austrian regiments of Loudon and Baden-Baden, and the company of grenadiers across the Grosser Spitzberg. The Prussian attack, supported by the fire of 60 guns, pushed back the Grenadier Regiment of the Observation Corps, and the the four regiments of musketeers. The Mühlberg was taken by the Prussians together with 42 cannons. Faced with an attack on the Grosser Spitzberg, Saltykov reformed his regiments in successive lines, the available space being too small to allow more than two regiments to form up facing the Mühlberg. A counter-attack failed to take the Mühlberg but slowed down the Prussians, who proceeded to bombard the Observation Corps putting them to flight together with the Grénadiers à Cheval.

Frederick now prepared to attack the Grosser Spitzberg from three sides, from the marsh of Els-Busch, from the north, and from the Kuhgrund in the south, where he assembled his cavalry under Seydlitz. The right attack was repulsed, but the second attack, that from the Kuhgrund, pushed the Russians back to within 150 yards of their principal battery, but fresh troops from the Judenberg arrived in the nick of time to fire point blank at the attackers. The third attack was made by Seydlitz's cavalry under a terrible artillery fire, and they were forced to retire with heavy losses. They were then charged, in turn, by the Austrian and Russian cavalry, accompanied by Cossacks, and were pushed back beyond the pond at Kunersdorf.

Saltykov next brought the whole of Fermor's 1st Division across the Grosser Spitzberg, chasing the Prussians into the Kuhgrund, while the troops occupying the Mühlberg were subjected to a violent artillery barrage. Then the Russians cleared the Kuhgrund at the point of the bayonet and retook the Mühlberg. Frederick fought in vain to stop the flight of his infantry, while Seydlitz's cavalry tried a final charge during which he was wounded, and his men broke. The Russian infantry came down from the heights, and the cavalry and cossacks spread out over the plain. On the Hühnerfluss a last stand by the Prussians was broken up by Cossacks who pursued their quarry along the paths between the ponds at Bischofsee. Frederick was only able to salvage 5,000 men and 50 guns from the rout. He had lost 19,000 men, most of his heavy artillery, and the recently-formed horse-artillery battery. The Russians lost some 2,600 killed and 10,000 wounded. At the end of the month after waiting in vain for Daun and his Austrians to come to his support, Saltykov withdrew back across the Oder, saving the badly shaken Prussian army from complete destruction.

A battle map is provided at the end of the book.

93. Prussia: Private, Dragoon-Regiment No. 6. (Schorlemer)

This regiment was taken into the Prussian service in 1717 from Saxony, reputedly exchanged by Frederick William I for a collection of porcelain, hence its nickname the '*Porzellan-Regiment*'. It was one of the 2 Dragoon regiments with 10 squadrons, the Bayreuth Dragoons being the other (see fig. 33). It was present at Zorndorf and heavily engaged at Kunersdorf.

94. Prussia: General Hans Joachim v. Zieten

General Zieten is shown here in the uniform of the regiment which he commanded from 1741–1786, Hussar-Regiment No. 2** also known as the *Leib-Husaren-Regiment*. This regiment distinguished itself at Prague, Zorndorf, Hochkirch, and Torgau. Zieten was born in Brandenburg in 1699. He joined the Prussian army, but his small size and quarrelsome nature led firstly to lack of promotion and eventually to cashiering in 1730. The following year, however, he was recalled and given command of the new squadron of the Life Hussars. In 1735 he was sent with 120 men to study the methods of the Imperial Hussars in their campaign on the Rhine, and on his return he was promoted major. Under his command the Prussian Hussars first made their reputation during the First Silesian War, but between 1745 and 1756 his peacetime clumsiness put him out of favour with the King. He returned to favour during the Seven Years War, during which as a General he was wounded at Kolin, and commanded a corps at Torgau. He died in 1786, a few months before his friend and sovereign.

95. Prussia: Gunner, Horse Artillery

Frederick the Great established the first properly organised horse artillery, one battery strong, at Landeshut in May 1759. By using light 6-pounders and mounting the 7-man crew on their own horses, the horse artillery could now keep up with the cavalry, travelling 'like the wind' wherever it was needed. Three months later the new battery was lost at Kunersdorf. A second battery was lost at Maxen, but a third, raised in 1760, survived the war. The Horse Artillery were dressed like the Foot Artillery except for the feather plumes in their hats, the gauntlets, and the cavalry boots and spurs.

96. Prussia: Grenadier Standard-Bearer, Infantry-Regiment No. 30. (Kannacher)

This regiment fought at Lobositz, Prague, Breslau, Leuthen and Hochkirch. The grenadiers, combined with those of Infantry-Regiment No. 7 to form the Grenadier-Battalion Lubath, were present at Kunersdorf. The

Free-Corporal shown here wears the waterproof canvas colour case wrapped around his body, and carries one of the company colours. General Warnery said of the Prussian Grenadiers at Prague that they were 'the only ones who did not open fire, but pressed home the attack at the bayonet point! After all they were Pomeranians . . . who are beyond doubt the best infantry in the world'.

97. Prussia: Grenadier Sergeant, Infantry-Regiment No. 20.* (Bornstedt)

This regiment fought at Kolin, Breslau, Leuthen, Olmütz, Hochkirch, Kunersdorf, Maxen, and Torgau. It had heavy losses at Hochkirch and Torgau. The Grenadiers served with those of Infantry-Regiment No. 5. to form Grenadier-Battalion Jung-Billerbeck.

98. Prussia: Grenadier Pioneer, Infantry-Regiment No. 17. (Manteuffel)

This 'good Pomeranian regiment' fought at Prague, Kolin and Torgau. The Grenadiers served at Kunersdorf with those of Infantry-Regiment No. 22 to form Grenadier-Battalion Tann. Pioneers wore leather aprons and carried axes and saws.

99. Prussia: Fusilier, Infantry-Regiment No. 35. (Fusilier). (Prinz Heinrich)**

This regiment fought at Reichenberg, Prague, Kolin, Breslau, Leuthen, Olmütz, Kunersdorf, Torgau and Freiberg. The Grenadiers served with those of Infantry-Regiment No. 36. The title of Fusiliers was given by Frederick to the 16 or so regiments raised after the conquest of Silesia. Drawn principally from newly acquired provinces where the men did not have the physical stature of the Pomeranians and Brandenburgers, Frederick gave them shorter muskets, and a revised form of Grenadier cap, which seems to have been based on one already worn by Saxon and Polish Grenadiers.

100. Prussia: Officer, Garrison-Regiment No. 5. (Jung-Sydow)

The Garrison-Regiments together with the companies of Garrison artillery were entrusted with the guard of fortresses and were composed of 'invalids and other officers who are too bad to appear in the field', nevertheless Frederick was forced to use them in his campaigns against the *Reichs* army and the Russians. Garrison-Regiment No. 5 fought at Kunersdorf, and its Grenadiers served with those of Garrison-Regiment No. 10. They wore blue coats, usually without lapels and blue waistcoats and breeches.

101. Prussia: Sergeant, Infantry-Regiment No. 41. (Fusilier). (Wied)**

Originally a Württemberg regiment, this regiment fought at Prague, Kolin, Leuthen, Hochkirch, Kunersdorf, Leignitz and Torgau. The Grenadiers served with those of Fusilier-Regiment No. 44 as Grenadier-Battalion Unruh.

102. Prussia: Pandour, Free-Corps v. Kleist

The Free-Corps v. Kleist was one of the largest of the irregular units consisting of infantry and cavalry. A battalion of infantry known as the 'green Croats' was raised by v. Kleist, who was himself *chef* the 'Green Hussars (see fig. 9), which included a detachment of Jägers. In 1762 they were present at Freiberg, when Prince Henry, with an army composed mostly of Free-Corps and Garrison regiments, beat the Austrians. At the end of the war, two battalions strong, they were disbanded, when the Free-Corps were tricked into laying down their arms. The officers were discharged and the men enrolled into the line or formed into Garrison-Regiments.

103. Prussia: Engineer Officer

Frederick inherited 45 military engineers, led by the Dutch Colonel v Walgrave, who, as a major-general, was in 1742 made *chef* of the new Pioneer-Regiment. In 1748 Walgrave fell into disgrace, was imprisoned for life in one of his own fortresses, and his regiment was converted to Fusiliers (No. 49), in which capacity it distinguished itself at Kunersdorf. The demoralized survivors of his Engineer corps, consisting of 5 chief engineers, 21 captains, 20 lieutenants, and 13 'conductors' were split up among the various Prussian fortresses. The officer shown here wears the cross of the *Pour le Mérite*, founded by Frederick in 1740, as a reward for anything from a good turnout on parade or a useful invention to gallantry in the field.

104. Prussia: Dragoon, Free-Corps v. Kleist

The 4 squadrons of Dragoons, raised by v. Kleist in February 1760, had increased by the end of the war to a regiment of 10 squadrons. The rank and file wore the fur cap, peculiar to this unit, but the officers and trumpeters wore tricornes. A squadron of Hussars, raised in 1760, had also increased to 10 squadrons by the end of the war, as had a body of curiously dressed 'Uhlans'.

105. Austria: Grenadier, Dragoon-Regiment Kolowrat

Raised in 1683, this regiment fought at Lobositz, where the Grenadiers particularly distinguished themselves, at Kolin, Breslau, Leuthen, and Hochkirch. At Kunersdorf, together with the Russian Arkhanguelgorod and Tobolosk Dragoon Regiments, it made a successful charge against the Prussian cavalry. It was also at Landeshut and Leignitz. There was one Grenadier Squadron of 2 companies to each regiment.

106. Austria: Private, Hussar-Regiment Kalnocky

Raised from the Hungarian National Militia, at the beginning of her reign, Maria Theresa had 14 regiments of Hussars by the time of the Seven

Years War, each of 6 squadrons of 200 men. The proprietors of the Hussar regiments were all Hungarian magnates. Raised in 1742, the Kalnocky Hussars fought at Kolin, Moys, Landeshut, and with distinction at Kunersdorf, where with the Austrian Dragoons and Russian Cavalry, the Austrian Hussars broke Seydlitz's squadrons. The Hussars were armed with a curved sabre, 2 pistols, and a carbine.

107. Austria: Pandour, Karlstädter Likaner Grenzregiment

In 1740 Maria Theresa raised bands of 'Croats' who, under Nadasti and Loudon, and the brutal partisan leaders Muetzel and Trenck, acquired a reputation for excellence in guerrilla warfare. Croats and Pandours were recruited from the Christian refugees settled along the borders with Turkey, where they formed 'Grenz' or frontier units for their own protection. By the time of the Seven Years War they had been organised into military districts, Karlstädter, Warasdiner, and Slavonian, each with a number of infantry battalions and hussar regiments. Raised in 1746, this regiment was one of the 4 Karlstädter regiments and fought at Prague, Moys, Schweidnitz, Kunersdorf, Landeshut, Leignitz, Glatz, and Chemnitz. Each infantry regiment consisted of 2 grenadier companies, 12 field companies, and 4 'home defence' companies, together with sharpshooting and artillery detachments.

108. Austria: Pandour, Slavonisch-Peterwardeiner Grenzregiment

Raised in 1747, this regiment was one of the 3 Slavonian infantry regiments, and fought at Kunersdorf, Maxen, Landeshut and Schweidnitz. The tactics of the Croats were to annoy and snipe at the enemy, very often at dawn and dusk, which they did to such good effect at Kolin that they totally upset Frederick's battle plan, and ultimately caused his defeat. Their tall felt cap was the origin of the 'shako' which later became so popular. Their dress which was based on their national costume was particularly practical especially the red cloak, and the light moccasins.

109. Austria: Musketeer, Free-Battalion Loudon

Raised in 1758 as a battalion of 4 companies, it was soon expanded to 2 battalions. The unit distinguished itself, in 1759, at Arnsdorf, and at Kunersdorf. It was also present at Landeshut and Leignitz. The green and red uniform worn by the Free-Battalion Loudon is said to have been chosen by Loudon in memory of his former service in the Russian army.

110. Russia: A Zaparogue Cossack

The Russian army during the Seven Years War could draw on the services of hordes of armed Cossacks, some of whom like the Don and Tchougouev Cossacks were formed into rudimentary regiments; at the accession of Elizabeth in 1741 the Don Cossacks could put 15,000 fully mounted men into the field. The word 'Cossack', of Turkish derivation meaning 'Free Warrior', was applied to a variety of peoples in the south of European Russia. Like the Croats, Pandours, and Hussars in the Austrian service, they occupied frontier settlements and acted as a barrier between Russia and her Turkish and Tartar neighbours. The figure here, taken from Viskovatov, is of a Zaparogue Cossack. These people for a long time lived an independent existence in the islands of the lower Dneiper, from whence they sallied out frequently on military expeditions. Their dress was not regularized in any way.

111. Russia: Hetman of Ukraine Cossacks

The Ukraine Cossacks, who won independence from Poland in the seventeenth century retained much of their old Polish customs and dress, and were the first of all the Cossacks to have anything like a regular uniform. The Hetman, or commander-in-chief, shown here is based on a figure of a Polish magnate in a painting by Bernardo Bellotto and carries in his hand the Hetman's staff of office. The third main type, the Don Cossacks, dressed according to fancy in kaftans, demi-kaftans, silk Turkish sashes, yellow red boots, fur trimmed caps with a cloth crown, and a variety of Russian, Turkish, Persian, and Tartar weapons.

112. Russia: Cornet of Cuirassiers of the Line

In 1756 there were 5 regiments of line cuirassiers each consisting of 5 squadrons of 2 companies, totalling some 796 all ranks. They were all dressed alike. Officers' ranks were distinguished by different sizes of gold lace, field officers having one row ¾ in. wide, and subalterns 2 rows ⅓ in. wide. The standards were carried by the cornets (*praportchnik*); a white standard was carried by the 1st Squadron and the remainder were blue; all bore the black double-headed eagle. The cuirasses were of the 1731 model with new ciphers for Elizabeth Petrovna. NCOs had ½ in. gold lace around their collars, and varying numbers of rings of gold lace around their cuffs, *viz.*: *Vahkmeister* 3, *Capitaines d'armes* and *Kvartrmistr* 2, and *Gefreit-Kapral* 1. The men's coats were trimmed with 1¼ in. red cloth. The hats had metal *calottes*. The officers were armed with swords and the men with swords and carbines. The 10 trumpeters as well as the kettledrummer, had buff swallows' nests and their coats and waistcoats were trimmed with red lace with yellow edges and crosses (see fig. 118).

113. Russia: Dragoon

In 1756 there were 29 regiments of Dragoons in the Russian army, all with geographical titles, each consisting of 1 squadron of Grenadiers and 5 squadrons of Dragoons. Each regiment had 2 3-pounders and 4 6-pounder mortars attached to it. Officers wore similar uniforms to the Dragoon shown here but with an officer's sash.

114. Russia: Horse Grenadier

In 1756 three Dragoon regiments Riazan, Kiev, and Novotroitz, were converted to Horse Grenadier Regiments. They were to consist of 5 squadrons of Grenadiers with the usual regimental artillery. They were dressed the same as the Grenadier companies of the Dragoons, but all, including the drummers wore grenadier caps.

115. Russia: Officer, Hussar Regiment Grouzinski

When Hussars were first created in the Russian army, in 1741, they were given a uniform inspired by the current Austrian pattern. Some regiments wore fur caps and some felt *mirlitons*. By the outbreak of the Seven Years War there were 8 Russian Hussar regiments, each of 5 squadrons. The trumpeters and kettle-drummers of Hussar regiments were dressed like the men with swallow's nests on the sleeves of the dolman and various coloured lace. The officers were armed with a sabre, and the men with a sabre, carbine and bayonet.

116. Russia: Sergeant, Grenadier Regiment

On 30 March 1756 the grenadier companies of the 3rd Battalion of every infantry regiment were amalgamated to form 4 Grenadier regiments. All ranks of these were dressed in the same way as the Grenadiers of a Musketeer regiment, the cap plate, and the flaps of the cartouche box and waist pouch, however, bore the Imperial eagle instead of the regimental badge.

117. Russia: Officer, Line Infantry

The Russian Line infantry consisted of 41 regiments, each of 3 battalions. When first raised they had borne the names of their Colonels, but in 1708–9, they were given territorial titles. Each battalion consisted of 4 Musketeer companies, and 1 Grenadier company The Russian army was the only one to have a system of rank badges for the officers, thus field officers wore gold and black sashes and gilt gorgets, while the subalterns wore yellow and black sashes and silver gorgets. Both types of gorget had the badge of the regiment within a trophy of flags in gold, and was hung from a red or sky-blue ribbon. The officer here wears a waist pouch, and is armed with a fusil and bayonet. The rank and file musketeers were generally as fig. 116, but with hats, and no waist pouches, instead of which a large black cartouche box, with brass corners, and a regimental plate was worn.

118. Russia: Drummer, Line Infantry

The Musketeer companies had drums and hautbois and the Grenadiers drums and fifes. The coats were laced with woollen braid to the Colonel's choice, Viskovatov showing the lace as having a pattern not unlike the design of the French royal livery lace. The drums had brass shells with the regimental badge applied in relief.

119. Russia: Musketeer, Observation Corps

Created in 1756, the Observation Corps consisted of a regiment of 4 battalions of Grenadiers, and 5 regiments of Musketeers of 4 battalions, each of 1 Grenadier and 3 Musketeer companies. Their dress was the same as that of the line regiments except that all ranks wore cavalry boots. The rank and file carried sabres, and instead of the territorial badges worn by the line regiments, they wore, on the caps and pouch flaps, a special badge consisting of the Imperial eagle in a halo of rays, with trophies and the Imperial cipher. The Grenadiers were dressed as those of the line, except for the boots, sabre, and special badges. The performançe of the Observation Corps at Kunersdorf left something to be desired.

120. Russia: Grenadier, Pandour Regiment No. 1

Two regiments of Pandours, after the Austrian Model, were raised in 1752. They consisted of 5 companies of Grenadiers and 15 of Musketeers. The latter had felt caps like those of the Grenadier shown here, but with a small double-headed eagle badge. The officers were armed with a sabre and 2 pistols, and the men with a sabre, 2 pistols, musket and bayonet. The drums and fifes were dressed as the rank and file, but with swallow's nests and red lace trimming.

121. Russia: Artilleryman

In 1757 the Russian artillery consisted of 2 regiments of Artillery, 3 companies of Siege Artillery, the Field Artillery of the Observation Corps, the Corps of 'Secret Howitzers', the Train, and the Pontoon Company. The 2 regiments consisted of 2 battalions each of 1 company of bombardiers and 4 of gunners. The bombardiers wore caps similar to the grenadier caps but lower, with different badges and with a brass crest made in the shape of falling feathers.

MONONGAHELA 9 JULY 1755

After 1748, in the absence of an international boundary agreement, France and Britain both tried to claim jurisdiction over the lands to the west of the existing British colonies, in particular the territory adjoining Pennsylvania and Virginia, with its rich fur trade, strategic waterways, and vast tracts of virgin land. A vital area was the so-called 'Forks of the Ohio', to which France, Virginia, and Pennsylvania all had direct access. If the British were to occupy it they would drive a wedge into the weak centre section of France's American colonies. If the French were to win the race they would put a stop to the fur trade and to British expansion in the west.

The French did in fact win when, in April 1754, they occupied the Forks and built Fort Duquesne. A small force of Virginians and British regulars, under Lieutenant-Colonel George Washington, endeavouring to set up an alternative outpost to the south, were forced to surrender their makeshift palisade, christened 'Fort Necessity' on 3 July, under terms which tacitly recognized French sovereignty over the area. This defeat forced the Virginians back across the Appalachian mountains, and did much to convince the British government that the threat of French dominion in North America was serious. As a result a force, consisting of the 44th and 48th Regiments together with a train of artillery, was sent to America, there to be joined by two new regiments, Shirley's and Pepperell's, to be raised entirely in America. It was felt that this force, with the help of some Provincial regiments, would be strong enough to retake the French advanced posts. Their commander, Major-General Edward Braddock, had orders to proceed to Virginia, with the 44th and 48th Regiments, and thence to lead an expedition of regulars and provincials against Fort Duquesne. At the same time Governors Shirley and Pepperell, with their new regiments, were to commence operations on the northern frontier, while Governor Lawrence, of Nova Scotia, was to seize the French outpost Fort Beauséjour, the key to Louisberg.

These plans were soon known to the French, and although neither country wanted war, 3,000 French regulars, under Baron de Dieskau, were assembled for service in America. Both countries had come to the conclusion that regulars were needed to prevent a disastrous tilt in the military balance of power, a significant departure, which was to make the next war different from all those that had preceded it.

At the end of April 1755, Admiral Boscawen sailed for America with a squadron of 13 sail, and on 8 June came up with the fleet carrying Dieskau and his six regiments. The British took two 64-gun ships, *L'Alcide* and *Le Lys*, capturing a number of French engineers and eight companies of infantry. In response the French recalled their Minister in London, and formal negotiations to preserve peace came to an end.

Throughout May Braddock's force assembled at Fort Cumberland, and at the beginning of June began its march towards Fort Duquesne. Its progress was extremely slow, however, and on 16 June Braddock pushed ahead with 1,400 officers and men, 10 guns and 30 waggons leaving the remainder to follow on at their best speed. By the evening of 8 July the advance guard lay on the banks of the Monongahela river, which once safely crossed, would bring them within eight miles of their objective.

Early the following morning Lieutenant-Colonel Thomas Gage moved off with a force of 300 regulars and two 6-pounders to secure the double crossing of the fords. Meanwhile however, Contrecoeur, the French commander at Fort Duquesne, sent out a force of 108 regulars, 146 Canadian Militia, and about 650 Indians under Captain de Beaujeu. Unknown to each other the two opposing forces were headed for a collision, which occurred about one o'clock when the British had safely negotiated the river and a ravine, and were passing through a large wood. In the initial confusion of the sudden encounter Beaujeu was wounded and many of the Canadian Militia fled, but the remaining French quickly split like a stream and passed to either flank of the British column, seizing some high ground nearby. Gradually Gage withdrew, abandoning his guns, and ran into a detachment sent forward by Braddock. There was now total confusion in the British ranks, as they tried in vain to form up and seize the hill by their unseen foe. After two hours of being shot down Braddock's men fled across the Monongahela, leaving their artillery and their dead to the victorious French and Indians who broke off their pursuit to busy themselves with scalping and looting. Two days later the remnants met up with the train, having buried their mortally wounded commander on the trail, and the whole force withdrew to Fort Cumberland, having lost 63 out of 86 officers and 914 out of 1,373 rank and file, killed and wounded. The French lost three officers, two rank and file, three militiamen, and 15 Indians, while their wounded amounted to some 16 men. Among Braddock's papers the French found the plans for impending British operations further north.

One of the significant consequences of this defeat was a quickened interest among British and American officers, in the modification of standard British tactics. The Americans had already been developing a new style of warfare, based on a mixture of Indian tactics, colonial improvisations, and the most useful of the manoeuvres laid down in the standard military manuals. The desire to keep the best of the old and to combine it with the new, led to the formation, in 1756 of the Royal American Regiment and the celebrated Rogers Rangers.

The year 1755 ended with the French firmly in control of the Forks of the Ohio, Fort Niagara, and Crown Point. Only in Nova Scotia, where the British had taken Forts Beauséjour and Gaspareau, had they suffered anything like a reverse.

1756 began with a change of command on both sides, Montcalm arriving with reinforcements for the French, and the Earl of Loudon to take command of the British forces. The British plans for the year were all frustrated, however by the French capture of Fort Oswego. Early in 1757 the French raided and destroyed Fort William Henry, the British base on Lake George, blocking possible British operations in the area, and forcing Loudon to look elsewhere. He decided on an expedition to Louisberg at the mouth of the St Lawrence river, but after Montcalm finally took Fort William Henry on 9 August, he was forced to abandon the expedition.

British fortunes were at their lowest ebb when, in 1758, the effects of William Pitt's new global strategy began to make themselves felt. Pitt made the conquest of Canada one of his main aims, and proposed the following plan to accomplish it. Amherst, with 13,000 men was sent to besiege Louisberg. Abercromby with 12,200, was to advance from Fort Edward, northwards into the Champlain Valley. Forbes, with 6,900 was to start a western offensive towards Fort Duquesne. All were successful except Abercromby who was repulsed at Ticonderoga with heavy losses. Louisberg fell to Amherst on 26 July, and Fort Duquesne was abandoned by its garrison in November. Thus for the first time since May 1754 the British were in control of the Forks of the Ohio.

PLATE DESCRIPTIONS

122. Great Britain: Grenadier Sergeant, 44th Regiment of Foot (Halkett)

Raised in 1741, this regiment fought at Monongahela, Ticonderoga, Fort Niagara and Montreal.

The unfortunate General Braddock seems to have made an effort to adapt the clothing and equipment of his troops to suit the conditions they were likely to encounter. He lightened their loads as much as possible by dispensing with shoulder and waist-belts and hangers; and to cope with the hot weather he ordered that bladders of 'thin leather' were to be provided for the troops to wear inside the crowns of their hats. The thick woollen waistcoats and breeches were replaced by 'Osnaburg', or coarse linen. Officers were to march with fusils and sergeants with muskets, bayonets and swords. The sergeant here wears a crimson sash with a narrow stripe of the facing colour, and a pouch on a narrow waist strap.

123. America: Colonel George Washington

In 1754 Washington was given command of the Virginia Militia, a unit in which only the officers appear to have worn uniform, the men wearing their own clothes. During Braddock's expedition the Virginian officers wore buckskin hunting shirts instead of their regimentals, which at that time were blue, with scarlet facings and silver lace, scarlet waistcoat, and blue breeches. By 1759 they were wearing blue, with scarlet facings and gold lace, and scarlet waistcoat and breeches, which agrees with Peale's portrait of Washington painted some years later. The rank and file still, we learn, had 'no uniforms nor do they affect any regularity'.

124. Great Britain: Officer, 48th Regiment of Foot (Webb)

Raised in 1740, this regiment fought on the Monongahela, Louisberg, Quebec, Martinique, and Havana. The coat worn here is based on a portrait of Major Gabriel Christie, who took part in Braddock's expedition with the 48th. He carries a fusil and wears light boots.

125. Canada: Private, Quebec Militia

The French colonists of New France constantly allied themselves with the various Indian tribes, but the best of their fighters were the half-breed settlers who combined the courage and determination of the white man with the cunning and agility of the redskin. From 1672 every able bodied man from 15 to 60 was eligible to serve in his parish Militia company. In 1758 just over 1,000 were called to arms although 4 times that number were employed in transporting supplies. They received no pay but were allowed to buy their musket at cost price at the end of their service. Although expert at forest fighting they were unreliable when used against regular troops as their initial behaviour at Monongahela proved. In 1759 the Militia companies were brigaded, each brigade being distinguished by the colour of their knitted woollen cap, or *Tuque*, red for Quebec, white for Trois Rivières, and blue for Montreal. This and the multi-coloured woollen sash worn by this figure were typical items of Canadian folk dress. The Militiaman's equipment consisted of musket, cartouche box, and tomahawk.

126. France: Fusilier, Compagnies Franches de la Marine

All the French colonies in America came under the Minister of Marine, and were garrisoned, from the seventeenth century onwards, by 'Independent Companies of the Navy'. By 1757 there were 40 such companies in Canada, mainly stationed in large fortified places, although some also served on the frontiers. These companies took part in most of the campaigns and actions of the war. In spite of their title they were, in fact, Colonial regulars, with officers recruited locally. The rank and file usually came from France with the intention of settling in the colonies when their service was over. In the western outposts difficulties of supply made the wearing of uniform impracticable, and a variety of clothing was worn; Bejeu, an officer of the Independent Companies, appearing at Monongahela clad only in breeches and a gorget. The Fusilier shown here wears a *bonnet de police*, regimental waistcoat, and Indian leggings. The regulation coat was grey-white with blue cuffs and lining, sergeants having 2 rows of gold lace round their cuffs. The drummers wore the King's 'small' livery, and the shells of the drums were painted blue with a scattering of gold *fleurs-de-lis*.

127. France: Colonial Artilleryman

In 1743 an artillery company of *cannoniers-bombardiers* was established at Louisberg, followed by several more companies in Canada in 1750. Twenty officers and men were sent out from France in 1757 to reinforce the Colonial artillerymen. Sergeants wore 2 stripes of silver lace round their cuffs and corporals 1; the drummers wore the King's 'small' livery.

128. Great Britain: Private, His Majesty's Independent Companies of American Rangers

'Rangers' were superior Militia raised among trappers and frontiersmen, who were adept in native warfare and woodcraft. The celebrated 'Rogers' Rangers' were formed in 1757 from the 'ranging companies' of the New Hampshire Provincial Regiment, but, although they were paid by the King, they were neither regulars nor provincials. Starting off with a company of 4 officers and 60 men the Rangers gradually increased to battalion strength including a company of Indians led by their own 'officers'. Between 1756 and 1759 Rogers waged constant guerrilla warfare against the French, penetrating enemy territory in the dead of winter, destroying settlements and Indian villages, and watching the progress of the new French Fort Carillon (Ticonderoga). Some of the engagements, were extremely bloody affairs, and in the celebrated 'Battle of the Snow-shoes', in the winter of 1758, 146 of Rogers' Rangers were killed. In 1757 Rogers formed a cadet company which trained officers from the regulars in the art of forest-fighting. An attempt was made to keep the clothing in each company uniform, some wearing green and others grey. The Indians wore their own dress. The Ranger here wears a green hunting shirt, and a Scottish bonnet; and has his snow-shoes slung across his back.

129. Great Britain: Private, Howe's Light Infantry

In order to fill the shortage of 'light troops' in North America the custom grew of selecting the 'most active resolute men' from each regular regiment and forming them into a body of 'Light Infantry'. Such a body was formed at Louisberg in 1758, and together with the Highlanders and the Rangers earned the title of the 'English Savages' from the French.

According to Amherst's and Wolfe's orders they were to crop their hats to turn them into caps, put the sleeves of their coats on their waistcoats, add large shoulder pieces or 'wings' to their coats, and 2 leather pockets for ball and flints. Their arms were a fusil, bayonet, cartouche box, powder horn and tomahawk. At first the Light Infantry took part in guerrilla operations with the Rangers, but in 1759, they led the army in its climb up to the Plains of Abraham. In 1760 the Light Infantry were returned to their regiments, but the impact they had made was to bear fruit some 10 years later when a light infantry company was added to each infantry battalion.

130. Great Britain: Private, 80th Regiment of Light-armed Foot (Gage)

Somewhat untypically the Duke of Cumberland told Loudon that 'till *Regular* officers with men that they trust, learn to beat the woods, and act as *Irregulars*, you will never gain any certain Intelligence of the Enemy'. Loudon took the hint and in September 1757 started sending 'gentlemen Volunteers' from the regulars to train with Rogers' Rangers, prior to setting up 2 ranging companies in each battalion. The exercise was cut short when Lieutenant-Colonel Gage, who had commanded the advance guard at Monongahela, offered to raise and clothe, at his own expense, a regiment of 500 Rangers, the costs to be reimbursed if the scheme met with Royal approval. The regiment was raised during the winter of 1757–8 and first saw service during the unsuccessful attack on Fort Ticonderoga. A sergeant and 4 men of the regiment were the first to enter the Fort during the successful expedition of 1759. The regiment was disbanded after the war.

QUEBEC 13 SEPTEMBER 1759

After the victories of 1758, conditions were now right for the long-delayed thrust into the heartland of Canada, the St Lawrence valley between Quebec and Montreal. Pitt's plan for 1759 was therefore simple in its objective. Amherst, with 11,000 men, including 6,000 regulars, was to move from the Hudson valley to the upper St Lawrence while Major-General James Wolfe with some 8,500 men was to advance from Louisberg up the St Lawrence River to Quebec, there to combine with Amherst in the conquest of Quebec or Montreal, or both.

The fleet bringing the new Naval commander, Admiral Saunders, and a large reinforcement of regulars arrived at Halifax on 30 April, and by 26 June the 1st Division of the fleet had completed the hazardous 400 mile journey up the St Lawrence to Quebec. Throughout July and August Wolfe bombarded the city of Quebec, in spite of two French attempts to burn the fleet with fireships, while he tried to find a way round its apparently impregnable defences. On 31 July the British made an abortive attack on the French defences below the town, losing some 450 men in the attempt. At length, urged on by his brigadiers, Wolfe decided to try an attack above the city, and during the last days of August ships and troops were quietly passed upstream. Eventually Wolfe found a possible landing site at the Anse de Foulon, and by the evening of 12 September a large part of the British army was embarked some six miles above it. While Bougainville and a mobile force patrolled the riverline above Quebec, Montcalm and Vaudreuil, apparently unaware of Wolfe's intentions, remained in the Beauport lines well below the city.

On the night of the 12/13 September Howe embarked 1,800 light infantry and dropped downstream. At the same time a demonstration off the Beauport shore held Montcalm's attention. Soon after 4 o'clock in the morning the first troops waded ashore at the Anse de Foulon, and began ascending the cliff path which led up to the Plains of Abraham and thence to Quebec. Before daylight 4,500 men had been landed with scarcely any loss.

As Wolfe formed his battle line, with his right on the cliff edge, Montcalm hurried his troops across the St Charles River, to place himself between the British and Quebec. Both sides were of equal strength, but whereas the British were all regulars, even the regular French regiments had recently been brought up to strength with local recruits. Montcalm formed in 3 columns opposite the main sector of the British line. At last, for the first time in North America, 'two disciplined armies were fighting each other in an open field, according to the European practice. This was what war was supposed to be'. It was soon over. At 10 o'clock Montcalm ordered the advance and 50 yards from the British line his columns fired off a ragged volley. The British held their fire until the last possible moment and when the smoke from their volleys had cleared the French were revealed pouring in disorder towards Quebec. Both Wolfe and Montcalm were mortally wounded in this short action. Vaudreuil, now in command of the French managed to withdraw his army to safety, and Quebec surrendered on 19 September.

Meanwhile Amherst, who had entered the abandoned Fort Ticonderoga on 26 July and had reached Crown Point on 4 August, was unable to push on further so that Pitt's plan for 1759 was not completely realized. On 8 September, 1760, Montreal was finally taken by the British and the conquest of Canada was complete.

France and Britain continued their naval warfare in the Caribbean and along the important seaways of the world, nor did the loss of Canada deprive them of Louisiana, and, there too, active French resistance, including intrigue with the Indians against the British continued. In fact even the Peace of Paris of 1763 came too late to prevent the outbreak of a serious Indian uprising, which was to last seven years. During this same period the British colonists also found themselves unhappily involved in an Anglo-American political crisis that opened the way to the American Revolution.

PLATE DESCRIPTIONS

131. Great Britain: Private, 78th Regiment of Foot (Fraser's Highlanders)

Raised at Inverness, in 1757, as the '2nd Highland Battalion of Foot' specifically for service in North America, by Lieutenant-Colonel Simon Fraser, son of the Lord Lovat who was executed after the 1745 rising, this regiment left for America at its full strength of 59 officers, 40 sergeants, 40 corporals, 20 drummers, and 1,000 privates. Another company was added in 1758. The 78th took part in the siege of Louisberg, and the capture of Quebec, and was disbanded in 1763, grants of land being awarded to those who wished to remain in Canada. As far as is known the 78th wore the 42nd, or 'Government' tartan. They were armed with muskets, bayonets, 'side pistols', and broadswords.

132. Great Britain: Major-General James Wolfe

James Wolfe entered the army at the age of 13½ years as a volunteer, in 1740, and after passing in succession through Colonel Wolfe's Marines, the 12th Foot, and the 4th Foot, he was promoted major in the 20th Foot, in 1749 becoming lieutenant-colonel in the same regiment at the age of 23. He commanded the 20th until 1758, when he was appointed colonel of one of the newly raised regiments, the 67th. In 1757 he was made quarter-master-general of the force sent on the abortive expedition to Rochefort, and in 1758 he was sent to America with the local rank of major-general in command of one of the brigades taking part in the siege of Louisberg. In 1759 he led the successful attack on Quebec and was killed at the moment of victory. At this date there was no regulation dress for general officers, and in this painting by J. S. C. Schaak, based on a sketch by one of Wolfe's ADCs he wears a plain red uniform with civilian top boots. Over his shoulder he carries an officer's fusil, the bayonet for which hangs in his Grenadier's waistbelt. He has no sash or sword, and round his arm he wears a black mourning band for his father.

133. Great Britain: Grenadier, 42nd Highlanders

Highland independent companies were raised in Scotland, in 1667, for police duties, and continued for many years. In May, 1735, new companies were raised, referred to as 'Our Northern troops alias Black Watch' presumably from their duties which included the suppression of cattle stealing and 'black mail'. At the same time they seem to have had a uniform tartan, which later became the 42nd or 'Government' tartan. In May 1740 the companies were formed into a regiment numbered 43rd, and in 1749 they became the 42nd Highlanders. The 42nd served at Martinique, Guadaloupe, Ticonderoga, Montreal and Havana. They were made a 'Royal' regiment in July 1758, and their facings were changed to blue. Highlanders, with their light shoes and clothing were the original light infantry in the British army, having it was considered by a contemporary many of the advantages which the Croats and Pandours had over the French. They were armed with muskets and bayonets, Highland or 'side' pistols, and a broadsword of native design The Grenadier shown here, after Morier, has an unusual type of fur grenadier cap which in 1768 became standard for the whole army, and his tartan has a red over-stripe. He has the buff facings worn by the regiment until it was made 'Royal' in 1758.

134. Great Britain: Grenadier, 22nd Regiment of Foot (Whitmore)

Raised in 1689, this regiment came out to America as part of the re-inforcement for the siege of Louisberg, in 1758, and its grenadiers fought at Quebec as part of the 'Louisberg Grenadiers'. It later saw service at Montreal, Dominica, Martinique, and Havana.

135. Great Britain: Pioneer, 28th Regiment of Foot (Bragge)

There were 10 Pioneers in each battalion of infantry, 1 per company, commanded by a corporal. They were the 'odd-job men' of the battalion, clearing obstacles on the march, digging trenches, making fascines, cutting fire-wood, when in camp, which according to Corporal Todd of the 12th Foot they sold to the officers, and helping to make up the cartridges before a battle or review. Hogarth's painting of the march of the Guards to Finchley shows a Pioneer in a red cloth cap turned up in the facing colour, and ornamented in front with a crossed saw and axe. The Pioneer here wears his waistcoat and apron. The 28th were raised in

1694 and served at Louisberg, Cape Breton, Quebec, Montreal and Havana.

136. Great Britain: Corporal, 15th Regiment of Foot (Amherst)

Raised in 1685 this regiment was present at Rochefort, Louisberg, Cape Breton, Quebec, Montreal, Martinique, and Havana. This figure of a corporal going through the 'Platoon Exercise' is taken from a drawing of Corporal Jones of the 13th Foot made by Lieutenant William Baillie in 1753.

137. Great Britain: Ensign, 43rd Regiment of Foot (Kennedy)

Raised in 1741, this regiment served in Nova Scotia and Maine, 1757–8, and was present at Quebec, Montreal, Martinique, and Havana. Each regiment of infantry had 2 colours, the King's, or First Colour, and the Regimental, or Second. The first had 'the Great Union throughout', and the second was of the facing colour with the union in the upper canton. Regiments with white facings, as here, bore a St George's cross on a white field. The number and rank of the regiment, within a wreath of roses and thistles, were to be painted in the centre of each colour, except in the case of regiments with royal or 'ancient' badges.

138. Great Britain: Officer, Royal Regiment of Artillery

The need for improved technical training led, in 1741, to the establishment of the Royal Military Academy, at Woolwich, the first military educational establishment in Britain. The cadets of the Academy, numbering some 40, formed the senior company of the Royal Regiment. Thanks to the Academy the British artillery soon acquired a high reputation for rapid and accurate fire. Minden was largely won through the efforts of Philips, Foy, and Macbean, and the efficacy of their fire; and at Warburg the manner in which the British guns galloped forward with the cavalry astonished the onlookers. In North America Artillery officers played an important part, supervising the guns, in the defence of frontier outposts, besieging fortresses like Louisberg and Quebec, and in the open field. Although the orders laid down that plain waistcoats were to be worn, portraits of Philips and other officers, show the laced waistcoat still in use.

139. Great Britain: Private, 60th or Royal American Regiment

After Braddock's defeat, a Swiss soldier of fortune, Prevost, successfully proposed the raising of a regiment of 4 battalions, in America, under a British Colonel, but with a sprinkling of foreign officers, and Henry Bouquet and Frederick Haldimand, Swiss officers of distinction, were chosen to command the first 2 battalions. A Royal Warrant of 4 March 1756 established the regiment at 4 battalions of 10 companies each, in all 4,400 men, who were eventually recruited in Massachusetts, New York, Pennsylvania, where there were 90,000 German settlers, Maryland and North Carolina. The regiment was first numbered 63rd but on the disbandment of Shirley's and Pepperell's became the 60th. The regiment served at Louisberg, Cape Breton and Quebec, where 2 battalions were present. They were also with Abercromby at Fort Ticonderoga, and with Forbes at Fort Duquesne in 1758. The buff coat lining, unusual in a regiment with blue facings is mentioned in an inspection report.

140. Great Britain: Royal Navy Seaman

In 1756 the Navy 'Slop' Office was formed to regulate the supply of 'slops' or clothing to seamen, who had no official uniform at that time, although bulk buying gave a semblance of one. The dress of the sailor shown here is based on a figure which appears in the background of Benjamin West's painting of the death of Wolfe. About this time, to judge from a painting by Morier of about 1760, the grey traditionally worn by British sailors was giving place to blue. The Royal Navy was much in evidence during the campaigns in North America, handling and serving guns ashore, crewing landing barges, and generally helping out over the extensive inland waterways. An officer and a party of seamen from H.M.S. *Norwich* formed part of Braddock's expeditionary force.

141. Great Britain: Vice-Admiral Sir Charles Saunders KB

Wolfe's Naval counterpart in the Quebec operation was a successful naval commander of the second rank, who after making the round-the-world voyage with Anson, in 1740, earned over £100,000 in prize money by the capture of 2 ships alone. During the Seven Years War he was second-in-command to Hawke, and commander-in-chief in the Mediterranean. He closed his successful career by becoming First Lord of the Admiralty. Wolfe thought him 'a zealous brave officer', and Walpole said of him: 'No man said less or deserved more.' A uniform for officers of the Royal Navy was established on 13 April 1748. Flag-officers, captains, commanders and lieutenants were to have a 'dress suite and a frock', while midshipmen had one outfit only. Admiral Saunders, in his portrait after Sir Joshua Reynolds, wears Flag-officers' undress with the ribbon of the Order of the Bath.

142. Great Britain: Royal Navy Lieutenant

James Cook, the celebrated navigator, rose from able seaman in 1755 to master in 1757. As master of the *Pembroke*, Cook explored and buoyed the difficult passages of the St Lawrence River for the British Fleet on its 400 mile journey up to Quebec. Although Cook did not become a lieutenant until 1768, we have used his portrait here as it is a good representation of a naval lieutenant of the time. He wears an undress coat with a laced waistcoat. The dress coat had white cuffs and no lapels.

143. France: Engineer Officer

The Corps of Engineers, formed of Royal Engineer officers, was originally part of the Artillery, but after 1759 it became a separate unit with its own school at Mézières. The uniform shown here is described by Mouillard as being worn up to that date. Afterwards a blue coat without lapels, with gold buttons, black cuffs with 5 buttons, and red collar, was worn with a red waistcoat and breeches.

144. France: Major-General the Marquis de Montcalm

Louis-Joseph, Marquis de Montcalm-Gozon, was born at Nîmes in 1712. He joined the Hainault Regiment, in 1727, as an ensign, and as a captain, in 1735, inherited his title. After service in Bohemia he became in 1743 Colonel of the Auxerrois Regiment, and 3 years later was wounded and taken prisoner at Picaenza. In February 1756 he was given the American command and was made major-general. He died of wounds at Quebec. In this portrait, as was frequently the custom, he wears rich civilian clothes, with a cuirass over his waistcoat. He also wears the Cross of the Order of St Louis.

145. France: Bombardier, Marine Artillery

French naval uniforms, established in 1756, have already been discussed in the author's *Uniforms of the American Revolution* but there were various units which came under the control of the Minister of Marine, which are of interest. The first Independent Companies of Naval Troops were formed by Cardinal Richelieu in 1662, and afterwards went through many changes in name and organisation while their task remained broadly the same, namely to serve aboard naval vessels as marines, and to supervise the art of Naval gunnery. The Artillery of the Marine c.1750 did not consider themselves as common soldiers and were not submitted to military exercises, but the Bombardiers of the corps, who were responsible for bombs and mortars, seem also to have been trained infantry who acted as grenadiers. They are seen drilling in the background of Joseph Vernet's painting of Toulon dated 1755, in the uniform shown here. Soon afterwards the colours were changed to blue coats with red facings, as shown in the same artist's painting of Bordeaux.

146. France: Drummer, French Infantry-Regiment La Reine

Raised in 1661, and numbered 24th, this regiment was part of the force sent to North America in 1755 under the command of Dieskau, but 4 companies were captured by the British fleet under Boscawen *en route*. The remainder was present at the Battle of Lake George, in 1755, when Dieskau was taken prisoner; the expedition to Fort William Henry in 1757; the defence of Ticonderoga in 1758; and at Quebec. The drummers of the regiment wore the red livery of the Queen of France.

147. France: Officer, French Infantry-Regiment Royal-Roussillon

Raised in 1661, and numbered 37th, this regiment fought at Ticonderoga, in 1758, and at Quebec. The officer here is wearing *bottines*, or leather gaiter-boots as worn by the French Dragoon regiments (see figs. 166 and 167).

148. France: Grenadier Corporal, French Infantry-Regiment Languedoc

Raised in 1672, and numbered 53rd, this regiment was part of Dieskau's reinforcement of 1755, and was present at the Battle of Lake George, 1755, Fort William Henry, 1757, Ticonderoga, 1758, and Quebec. The ranks of French infantry NCOs were indicated by lace loops on the buttonholes of the cuff, and by a lace edging to the top of the cuff. Corporals had 3 loops in yellow or white worsted, and lance-corporals a ring of yellow or white worsted lace round the cuff. In some cases, as in the Regiment *Béarn*, the loops were laid on horizontally instead of vertically, sometimes they had fringed and sometimes pointed ends.

149. France: Fusilier, French Infantry-Regiment Guyenne

Raised in 1684 and numbered 68th, this regiment was part of Dieskau's reinforcement, and fought at the Battle of Lake George, 1755, Oswego, 1756, Fort William Henry, 1757, Ticonderoga, 1758, and Quebec. It was amalgamated with the 16th (Dauphin) in 1762. The Fusilier shown here wears Indian-style leggings, and carries a powder horn and tomahawk. He also wears his waist-belt over the right shoulder, as this was increasingly found to be more practical by both armies until eventually waist-belts were done away with completely, and the bayonet and sword were carried on a shoulder belt.

150. France: Ensign, French Infantry-Regiment La Sarre

Raised in 1661, and numbered 34th, this regiment was part of Dieskau's reinforcement of 1755, but lost 4 companies to the Royal Navy *en route*. The remainder was present at Oswego, 1756, Fort William Henry, 1757, Ticonderoga, 1758, and Quebec.

151. France: Grenadier, French Infantry-Regiment Béarn

Raised in 1684 and numbered 72, this regiment was as Oswego in 1756, Ticonderoga, 1758, and Quebec. It was disbanded in 1762.

EMSDORFF

The Russian victories at Zullichau and Kunersdorf, and the subsequent Austrian successes at Maxen and Meissen left Frederick, at the end of 1759, in an apparently hopeless position. The French also had been far from victorious. The Allied victory at Minden had driven them out of Westphalia and Hesse, while Hawke's notable victory at Quiberon Bay, on 20 November, had ruined any chances France may have had of invading England and of regaining her lost colonies. France was inclined to seek peace, but Prussia was determined to continue the war.

The end of the invasion threat released fresh British troops – the so-called 'Glorious Reinforcement' – for service in Germany and for his next campaign Prince Ferdinand had at his disposal 33 squadrons and 21 battalions of British troops totalling some 22,000 men. His other contingents, with the exception of the Prussian, which was needed nearer home, were all increased bringing his total strength to 80,000 men.

On 10 July 1760 the first part of the campaign in Hesse came to an end when de Broglie and Saint-Germain combined to turn Ferdinand's right wing at Corbach, but failed to follow up their success. Outnumbered, and with both flanks exposed, Ferdinand decided to strike out at the French supply base at Marburg, and detached the Erbprinz with a force of 6 Hanoverian and Hessian battalions, some Hanoverian Jägers, the Hanoverian Luckner Hussars, and the newly-arrived British 15th Light Dragoons.

On the evening of 15 July the Erbprinz learned of a French force of five battalions and a regiment of Hussars, under Glaubitz near the village of Emsdorff. The next morning he set out to reconnoitre, and found them lying north-east with their left on Emsdorff and the Burgholz forest and their right on Erksdorf, which they had left unoccupied. The Erbprinz and General Luckner rode to within half a mile of their camp without seeing any sign of outposts, or guards, so, placing one of his battalions, Luckner's Hussars, and the 15th Light Dragoons to close the trap in front of Erksdorf, he led the remainder of his force through the forest and surprised the French camp. Their Hussars promptly fled, but the infantry rushed to their arms. Before they could form up properly Luckner's horsemen charged in from the right, and the whole force abandoned its camp and withdrew in a south-westerly direction through the woods towards Langenstein where they rallied. Luckner's infantry battalion now by-passed them, blocking their way to the east, while the 15th Light Dragoons headed them back from the Amöneberg road to the south-west. The French next tried to escape southwards through the woods only to be blocked yet again. Eventually they broke out to the west on to an open plain where they were promptly charged by the 15th Light Dragoons who captured one battalion complete. Continuing their retreat the French reached a wood near Schweinsberg, halted and fronted once more, but were soon surrounded by the Erbprinz who summoned Glaubitz to surrender. The allies took 1,655 prisoners, numerous colours, six guns and all the baggage. The Erbprinz decided to abandon the raid on Marburg and withdrew.

'This gallant little affair' showed once again the superiority, man for man, of the allies over the French, but otherwise did little for the general furtherance of the campaign.

PLATE DESCRIPTIONS

152. Great Britain: Hornist, 15th Light Dragoons (Eliott's)

153. Great Britain: Private, 15th Light Dragoons (Eliott's)

On 17 March 1759 a regiment of light dragoons was raised by Colonel George Augustus Eliott, consisting of 18 sergeants, 18 corporals, 12 drums, 6 hautbois, and 360 privates, in 6 troops. The lieutenant-colonel was Lord Pembroke. The regiment had been in existence only a few months when it was sent to Germany, and within weeks of its arrival had become famous. Their casualties at Emsdorf were so heavy (125 men and 168 horses) that they had to be sent back to Hanover to reorganize. As Fortescue writes, 'It was the Fifteenth alone, a young regiment which had never been under fire, which thrice charged five times its numbers of French infantry and rode through them; . . . Their gallantry indeed was the amazement of the whole army'. The Light Dragoons were distinguished by their curiously shaped caps, and light 'jockey' boots. They were armed with carbine, bayonet, pistol, and sword, and carried entrenching tools in their holsters. Instead of housings and holster covers they had a one piece cloth 'shabraque' like those used by the Hussars, and light saddles. They had special training in horsemanship and in firing from the saddle. The drummers of the 15th wore green liveries with hanging sleeves, heavily laced, and their drums had red hoops and green fronts emblazoned with the Colonel's arms. The French hornist, shown by Morier in a painting at Wilton House had a simpler green uniform; he probably took the place of the established hautbois. In 1764 drums were abolished for light cavalry, the trumpeter from each troop, when dismounted forming a band of 2 French horns, 2 clarinets, 2 bassoons, with the addition of 1 fife.

154. France: Fusilier, German Infantry-Regiment Royal-Bavière

This regiment which was raised in 1709 by the Elector of Bavaria, for French service was numbered 86. It provided 2 of Glaubitz's 5 battalions at Emsdorff, the other 3 belonging to the German Infantry-Regiment Anhalt (No. 46), who wore blue with straw facings collars, cuffs, lapels, turnbacks, waistcoats and breeches.

155. France: Private, Hussar-Regiment Berchény

Prior to 1720 the Hungarian Count Ladislaw de Berchény obtained the King of France's permission to recruit in Wallachia, then a Turkish province, armed with a letter to the French Ambassador at the Sublime Porte. He soon returned with the first 100 officers and men who formed the nucleus of the Berchény Hussars, stationed in Alsace. The second French regiment of Hussars, they were numbered 59th in the list of cavalry regiments. During the wars of the Polish and Austrian successions the new regiment acquired a formidable reputation, its Colonel eventually becoming Inspector-General of Hussars and a Marshal of France. Gradually, while the officers remained Hungarian, the men became increasingly natives of Alsace, until, in 1760 Frenchmen, provided their usual tongue was German, were permitted to serve in Hussar regiments. The conduct of the Berchény Hussars at Emsdorff does not seem to have lived up to their glorious traditions.

WARBURG 31 JULY 1760

After Emsdorff, Ferdinand broke contact with the French, by-passing his main arsenal at Cassel. On 29 July he learned that Saint-Germain's successor Du Muy was marching towards Warburg, on the River Diemel, in an attempt to turn his right flank. They outnumbered a force of 22 squadrons, 23 battalions, and 26 guns commanded by the Erbprinz and Spörckern, which was already in the vicinity. Eight miles to his left lay Cassel, with its weak garrison and poor defences. In this dangerous predicament Ferdinand decided to throw all his strength into supporting his detached force near Warburg.

By the morning of 30 July the French were encamped behind the ridge which runs north-west from Warburg, with their right on the town and their left where the ridge begins to fall away towards the isolated village of Ossendorf. Behind them rose the Hein Berg, an isolated feature surmounted by a watchtower. The Erbprinz and Spörckern decided to attack the French left, moving by night to a position in the area Ossendorf-Menne. Ferdinand, however, ordered the attack to be delayed until the main army had crossed the Diemel, which it was expected to do early the following morning. The morning of the 31st was misty, however, and Ferdinand, at 7 o'clock, ordered the advance to begin. At half-past nine, the mist cleared partially to reveal the main army, advancing from Liebenau, but not apparently the Erbprinz and Spörckern, to the French. The right-hand column emerging from Ossendorff went straight for the Hein Berg, which was unoccupied, and soon their regimental guns were firing on the French rear. Du Muy immediately hurled two of the four infantry brigades of his left flank into a counter-attack, and bitter fighting followed during which the British Grenadiers distinguished themselves. The Allied artillery came into action, and the left-hand column from Menne stormed forward, overrunning the French guns and taking the counter-attacking infantry in the rear. The left of the French line broke, and streamed back to Warburg with the British Royal Dragoons in hot pursuit.

Du Muy ordered the two infantry brigades on his right flank to cover the withdrawal of the army across the Diemel, and, as the cavalry which formed the centre of the line were about to conform, the British cavalry, 22 squadrons strong, accompanied by artillery swept down on them. As the main army had struggled forward, delayed by bogs and corn fields, Ferdinand had ordered the Marquis of Granby to take them forward and charge Du Muy's centre. Forming into lines as they went, they trotted and cantered onwards, led by Granby, 'his bald head shining in the sun', at the head of the Royal Horse Guards, arriving in time to catch the French cavalry in the act of retiring. Two brigades of French cavalry made a vain counter-attack, and soon joined their fellows fleeing towards the Diemel, where the Allied artillery completed their rout.

The Battle of Warburg, successful though it was, cannot be counted as Ferdinand's master stroke. All the heavy fighting was done by Spörckern's infantry which included Keith's and Campbell's Highlanders and the two British Grenadier battalions, and the British cavalry only arrived when the battle had been won; nevertheless their intervention was decisive, and Warburg has every right to be considered as one of the great cavalry actions, and one in which Granby restored the reputation of the British cavalry, so sadly tarnished by Lord George Sackville at Minden.

The Allies lost some 1,500 men out of 24,000 and the French out of 21,500 lost some 8,000 men.

PLATE DESCRIPTIONS

156. Great Britain: Private, 1st Dragoon Guards (Bland)

Raised in 1685, the regiment became the King's Own Regiment of Horse in 1714. In Ddcember 1746 the first three regiments of Horse were for reasons of economy 'reduced to the quality of Dragoon', with the extra title of 'Guard' added as a compensation, losing their crossbelts and being provided with drums and hautbois instead of trumpets. Its establishment in 1750 was 9 troops with 2 drummers and 1 hautbois to each. In 1756 it was augmented by the addition of a 'light' troop (see fig. 160). The regiment fought at Minden; Corbach, where it charged with the 3rd Dragoon Guards in an attempt to save some Allied guns losing 66 men out of 90; Vellinghausen; Wilhelmstahl, and other engagements in Germany. The 1751 Warrant laid down that Dragoon Guards should wear lapelled coats with slit cuffs, and shoulder knots of white or yellow worsted on the right shoulder. The coats were to have long pockets, and the button-holes were to be of narrow yellow or white lace set on in twos and threes. Sergeants were to have gold or silver lace and shoulder knots and corporals gold and silver lace with a yellow or white silk shoulder-knot. Their arms were sword, carbine, and pistols.

157. Great Britain: Lieutenant-General The Marquis of Granby

John Manners, Marquis of Granby was a 'generous, genial character much loved by his troops and popular with the public'. Born in 1721 he became Colonel of the Royal Horse Guards and in 1755 a major-general. In August 1759 he succeeded the disgraced Sackville as the Commander-in-Chief of the British Army in Germany and led it throughout the campaign, showing much gallantry. He became Master-General of the Ordnance in 1763 and Commander-in-Chief in 1766. He died at Scarborough of gout, in 1770. In this portrait after Edward Penny he wears the uniform of his Regiment, the 'Blues' which was raised in 1661. It was present at Minden, Warburg, Vellinghausen, and Wilhelmstahl.

158. Great Britain: Sergeant, 3rd Dragoon Guards (Howard)

Raised in 1685 and converted from the 4th Horse to the 3rd Dragoon Guards, in 1746, the regiment was present at Minden, Corbach, where it charged with the 1st King's Dragoon Guards, Vellinghausen, Wilhelmstahl and other engagements in Germany. The Light Troop raised in 1756 took part in the expeditions to St Malo and Cherbourg in 1758.

159. Great Britain: Private, 6th Dragoons (Cholmondeley)

Raised in 1689. The Light Troop, added in 1755, saw service at St Malo and Cherbourg, and the regiment was at Minden, Warburg, Klosterkampen, Vellinghausen and Wilhelmstahl. According to the 1751 Warrant, the Dragoons were to have coats without lapels, either with buttons and holes on both sides or buttons on one side and holes on the other, otherwise their laced buttonholes, pockets, shoulder-knots and rank distinctions were the same as the Dragoon Guards. The 'camp necessaries' of Dragoons, in 1760, consisted of haversacks, canteens with 'hempen strap', stable frocks, and ash picket poles, 4 ft 4 in. long, with a 'staple and ring at top'.

160. Great Britain: Private, Light Troop, 11th Dragoons (Ancram)

Raised in July 1715, this regiment part of the 'glorious reinforcement' was present at Warburg, Vellinghausen, Wilhelmstahl, the siege of Cassel, and other engagements in Germany. Following the valuable services of Kingston's Light Horse during the 1745 rebellion, and of the Duke of Cumberland's Light Dragoons in Flanders, it was decided to raise a body of Light Cavalry in the British army, and at the end of 1755 a troop of 'Light Dragoons' 3 officers and 65 men strong, was added to 3 Dragoon Guard Regiments (1st, 2nd, and 3rd) and 8 Dragoon Regiments (1st, 2nd, 3rd, 4th, 6th, 7th, 10th and 11th). The men were to wear 'jacked leather caps' and 'light jockey boots' and their carbine belts were altered to allow them to fire on horseback. Some of them first saw service on the expeditions to St Malo and Cherbourg. After the Seven Years War the Light Troops were disbanded.

161. Great Britain: Officer, Light Troop, 2nd Dragoons (Campbell)

In 1681 various troops of Scots dragoons were regimented and in 1736 they first appear as the Royal North British Dragoons commonly known as the Scots Greys. The Light Troop of the regiment, which wore a cut down cloth fronted mitre cap similar to the Grenadier caps worn by the rest of the regiment, saw action at St Malo and Cherbourg. The regiment was present at Minden, Warburg, Vellinghausen, Wilhelmstahl, and other engagements in Germany. The regiment seems to have worn hats on service in Germany although they apparently had their Grenadier caps with them as the 'Orderly men to Lord Granby's' were ordered to wear them in oilskin covers in bad weather.

162. Hanover: Private, Cavalry-Regiment Bremer

There were 8 regiments of Horse in the Hanoverian army, each consisting of 2 squadrons of 3 companies, totalling some 350 men. There were also 4 Dragoon regiments of 4 squadrons of 2 companies each. They had their own grenadiers who wore Prussian-style Grenadier caps. Whereas the Horse wore white coats, with differing regimental facings, without lapels, the Dragoons had lapels, and 3 of the regiments, Breidenbach, Reden, and Wülfingen had in addition shoulder knots of the facing colour on the right shoulder. The Dragoons were armed with carbine and bayonet, pistol and sword.

163. Hessen-Cassel: Private of Dragoons

There were 2 regiments of Dragoons in the Hessen-Cassel service, each of 4 squadrons, totalling 662 per regiment. This figure is based on Morier's painting of a private of the Dragoon-Regiment König, c.1748. As far as we know the uniforms had not changed by the time of the Seven Years War.

164. Hanover: Private, Luckner's Hussars

Ferdinand's army contained a sizeable proportion of light troops, who were mainly recruited from foreign deserters and were not conspicuous for their discipline. Luckner's hussars in their colourful uniforms were easy to recognize, and soon became a familiar sight to the French and their allies. They took part in the Emsdorff operation where they aroused Lord Pembroke's anger by the speed with which they pounced on the abandoned French camp and 'got everything' without having done much to earn their plunder. Raised in 1757 from men in the Dutch service, as 1 company of 54 men, they were increased over the years until, in 1760, they had 4 squadrons. They had 2 uniforms, the first green, with yellow lace, and red breeches, and the second as shown here.

165. Hanover: Private, Freitag's Jägercorps

Another celebrated body of light troops, in Prince Ferdinand's army, this corps was originally raised in May 1757 by Graf v. Schülenberg, and consisted at first of 2 mounted and infantry companies. It was gradually augmented until, in 1759, when Lieutenant-Colonel Freitag succeeded to the command, it consisted of 3 brigades, each of 1 Mounted and 2 Infantry companies and 1 4-pounder gun. In 1760, each brigade was increased by one Mounted company. The mounted troops wore similar uniforms to the infantry, but with buff leather equipment and breeches, cavalry boots, carbines and swords.

166. France: Private Dragoon-Regiment Royal

167. France: Private Dragoon-Regiment Thiange

In 1667 the numerous companies of dragoons then existing were regimented and in the following year were placed under the control of a *Colonel-Général* and *Méstre-de-Camp* of Dragoons. At the outbreak of the Seven Years War there were 16 regiments each with 16 companies of 40 men, of which the Royal Regiment, raised in 1656 was numbered 3rd, and Thiange, raised in 1676, 14th. The Dragoons were, as was usual with most European countries, dressed and equipped like mounted infantry. The characteristic Dragoon bonnet, shown here was sometimes replaced on service with a hat trimmed with false silver lace. The coats were without lapels and had laced buttonholes like the British, and in 1750 leather boot-gaiters called '*bottines*' were introduced for all ranks. The Dragoons were armed with musket and bayonet, pistol, and sword; and carried either an axe, a saw, or a bill-hook in place of the off-side holster.

168. France: Sergeant, French Infantry-Regiment Bourbonnois

Raised in 1661 this regiment was one of the *Petit Vieux Corps* and was numbered 8th. Sergeants wore either 3 gold or silver lace loops on their cuffs or a border of gold or silver lace, as shown here.

169. France: Ensign, French Infantry-Regiment La Couronne

Raised in 1661 this regiment was numbered 28th. The ensign carrying one of the *drapeaux d'ordonnance* of the regiment wears a similar uniform to the men, except for his gorget and elaborately laced waistcoat.

170. France: Grenadier, French Infantry-Regiment Latour du Pin

Raised in 1661, and one of the *Petit Vieux Corps*, this regiment was numbered 7th. In summer it was frequently the custom for the troops to discard either their coat, or, as shown here, their waistcoat.

171. France: Hussar Private, Chasseurs de Fischer

172. France: Grenadier, Chasseurs de Fischer

During the siege of Prague, in 1742, an officer's valet, by the name of Fischer, made a reputation for himself guiding the other valets in and out of the islands of the Moldau river, to pasture the officers' horses, and they were eventually put under his command by Marshal Saxe. In November 1743 they were formed into a free-company of *chasseurs*, consisting of 40 horse and 60 foot. At the Battle of Lauffeld, in July 1746, the corps served as escort to the King, and as a reward Fischer was made Brigadier of Cavalry, and allowed to retain 4 companies at the Peace. During the Seven Years War the corps had 8 companies of 150 men each, half of whom were mounted but when Fischer died in 1761 it had increased to 3,600 men consisting of 2 battalions of infantry of 8 Fusilier, 1 Grenadier and 1 Chasseur company, 4 squadrons of Hussars, or *Chasseurs à Cheval*, 4 squadrons of Dragoons, and 2 artillery pieces.

EPILOGUE

After the disastrous events of 1759, Frederick managed to gain a victory over the Austrians, under Loudon, at Leignitz, in August of the following year, but Berlin was attacked and partially burnt by a Russo-Austrian force. In November, 1760, Frederick again attacked the Austrians, this time under Daun, at Torgau, ultimately with success although with considerable loss. In 1761, Frederick's fortunes were at their lowest ebb. He was for some time in considerable danger in the camp at Bunzelwitz, but was saved by a lack of co-operation among his enemies, the Russians marching away. Loudon, however, took the town of Schweidnitz, and the Russians captured Kolberg, in Pomerania.

The death of the Empress Elizabeth of Russia, in January 1762 altered the whole situation dramatically. The new Russian ruler, Peter III, abandoned the Austrian alliance, and even offered to come to Frederick's aid with a corps of 20,000 men. The dethronement and murder of Peter III put an end to this offer, but the new Empress Catherine refrained from entering the war on Austria's side. Daun was obliged to retire, and Schweidnitz was recaptured by the Prussians. In addition, Prince Henry of Prussia with a force largely composed of second-class troops fought a victorious action at Freiberg which brought the campaign to a close. Austria, left unsupported at the end of 1762 was forced to conclude peace with Prussia, which was signed at Hubertusburg on 15 February 1763, on the general basis of the *status quo ante bellum* thus leaving Prussia in possession of Silesia.

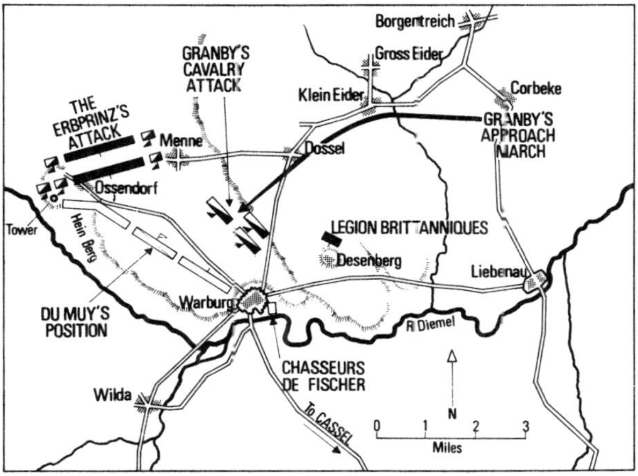

Warburg 31 July 1760

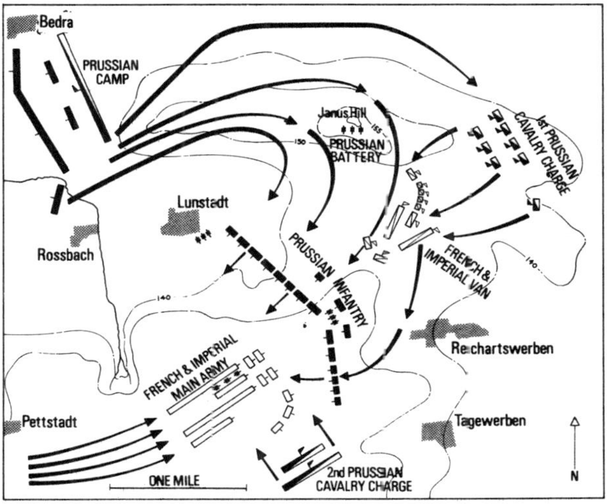

Rossbach 5 November 1757

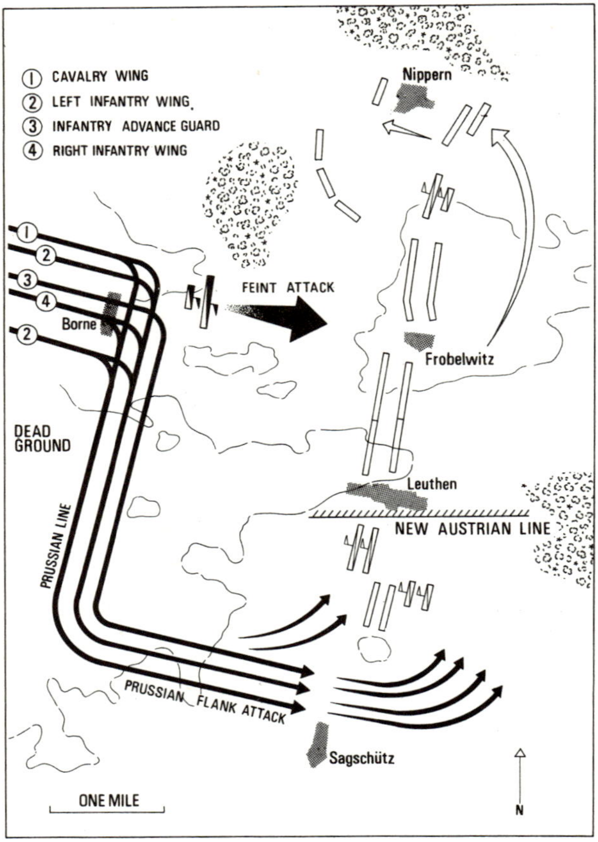

Leuthen 5 December 1757

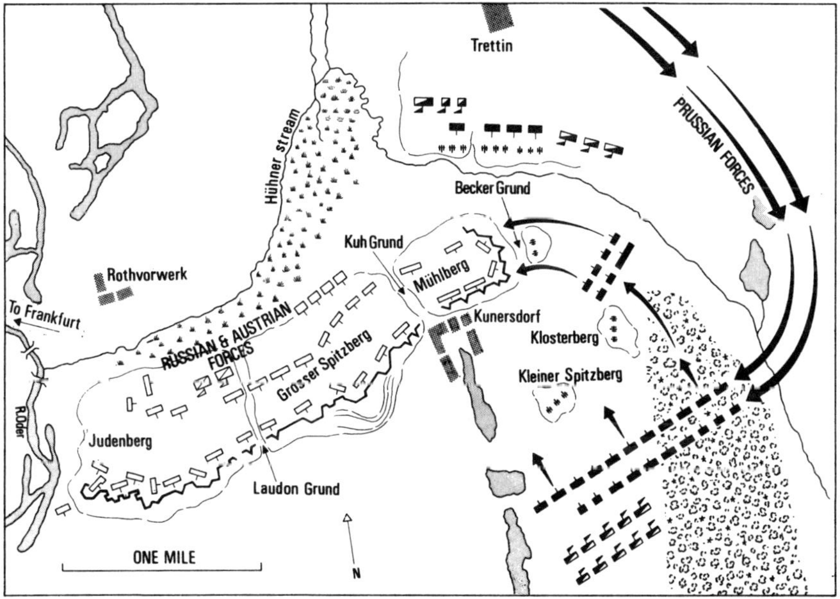

Kunersdorf 12 August 1759

APPENDIX

ORDERS OF BATTLE

I. ROSSBACH

1. Prussian Order of Battle

1. *Infantry and Artillery*

		Bns
1.	Grenadier-Battalion,	
	Wedell (1/23)	1
2.	Kremzow (17/22)	1
3.	Ramin (19/25)	1
4.	Field-Infantry-Regiment,	
	Margraf Karl (19)	2
5.	Meyerinck (26)	2
6.	II/III Garde (15)	2
7.	Retzow (6)	1
8.	Itzenplitz (13)	2
9.	Forcade (23)	2
10.	Kleist (27)	2
11.	Alt-Braunschweig (5)	2
12.	Grenadier-Battalion,	
	Jung-Billerbeck (5/20)	1
13.	Lubath (7/30)	1
14.	Field-Infantry-Regiment,	
	Winterfeldt (1)	2
15.	Golz (24)	2
16.	Hülsen (21)	1
17.	Grenadier-Battalion,	
	Finck (13/26)	1
18.	Free-Battalion Mayr (2)	1
19.	Artillery	
		27

2. *Cavalry*

		Sqns
20. Cuirassier-Regiment,		
	No. 13 Garde du Corps	3
21.	No. 10 Gensd'armes	5
22.	No. 8 Rochow	5
23.	No. 7 Dreisen	5
24.	No. 3 Leib	5
25. Dragoon-Regiment,		
	No. Meinicke	5
26.	No. 4 Czettritz	5
27. Hussar-Regiment,		
	No. 1 Czekeln	10
28.	No. 7 Malachowski	2
		45

3. *Formation*

*First Line** – 20–22, 2–12, 23, 24
Refused on flanks of infantry of first line – 1, 13
Second Line – 25, 14–17, 26
Reserve – 27, 18, 28

I. ROSSBACH

2. French and Imperial (Reichs) Order of Battle

1. *Infantry and Artillery*

		Bns
1. French Infantry-Regiment,		
	No. 4 Piémont	4
2.	No. 21 Saint-Chamond	2
3.	No. 57 Cossé-Brissac	2
4. German Infantry-Regiment,		
	No. 92 Deux-ponts	2

*Reading from right of the line to the left in all cases.

		Bns
5.	French Infantry-Regiment,	
	No. 37 Royal Roussillon	2
6.	Swiss Infantry-Regiment,	
	No. 51 Reding	2
7.	No. 63 Planta	2
8.	French Infantry-Regiment,	
	No. 47 Royal Artillerie	1
9.	No. 11 Mailly	4
10.	German Infantry-Regiment,	
	No. 66 La Marck	2
11.	Royal-Pologne	1
12.	Saint-Germain	1
13.	Swiss Infantry-Regiment,	
	No. 52 Castellas	2
14.	No. 87 Salis	2
15.	No. 55 Wittemer	2
16.	No. 77 Diesbach	2
17.	French Infantry-Regiment,	
	No. 14 Poitu	2
18.	No. 61 Provence	2
19.	Montbazon	2
20.	No. 41 Beauvoisis	2
21.	Kreis Infantry-Regiment,	
	Würzburg 'Blue' Infantry-Regiment	2
22.	Hessen-Darmstadt	1
23.	Varell	2
24.	Trier	2
25.	Kreis Artillery	0
26.	Kreis Infantry-Regiment, Cronegk	2
27.	Ferentheil	2
28.	French Infantry-Regiment,	
	No. 38 Condé	2
29.	No. 18 Touraine	2
30.	No. 6 La Marine	4
31.	Austrian Light Troops, Croats	3
		63

32.	Austrian Cuirassier-Regiment, Brettlach	7
33.	Kreis Cuirassier-Regiment, Prinz Friederich	3
34.	Hohenzollern	3
35.	Kreis Dragoon-Regiment, Württemberg	2
36.	French Cavalry-Regiment, No. 14 La Reine	2
37.	No. 37 Bourbon-Busset	2
38.	No. 56 Fitz-James	2
39.	No. 3 Commissaire-Général	2
40.	No. 69 Voluntaires Liègois	2
41.	No. 22 Bourbon	2
42.	Austrian Cuirassier-Regiment, Trautmannsdorff	7
43.	Kreis Cuirassier-Regiment, Brandenburg-Bayreuth	5
44.	Dragoon-Regiment, Brandenburg-Anspach	5
45.	French Cavalry-Regiment, No. 25 Penthiévre	2
46.	No. 2 Mêstre-du-Camp Général	2
47.	No. 7 Cuirassiers du Roi	2
48.	No. 19 Berry	2
49.	No. 58 Escars	2
50.	No. 18 Aquitaine	2
51.	No. 4 Royale	2
52.	No. 9 Royal-Roussillon	2
53.	No. 1 Colonel-Générale	2
54.	No. 21 Condé	2
55.	No. 27 Poli	2
56.	Austrian Hussar-Regiment, Szecheny	6
57.	French Hussar-Regiment, No. 69 Nassau-Saarbrück	2
58.	Dragoon-Regiment, No. 13 Apchon	4

59. Austrian Hussars	2
60.	1
	81

3. *Formation*

1. *Prince v. Hildburghausen and Prince de Soubise*
   First Line – 32–35, 1–7, 36–41
   Reserve Artillery – 8
   Second Line – 42–44, 9–16
   Reserve Corps – 45–47, 17–20, 48, 49
   Reichs Army – 21–27
2. *Saint-Germain*
   First Line – 50, 51, 28, 29, 52, 53
   Second Line – 54, 30, 55
3. *v. Loudon*
   First Line – 59, 31, 60
   Light Cavalry – 56–58

2. LEUTHEN

1. **Prussian Order of Battle** NB. * = present at Rossbach

1. *Infantry and Artillery*

		Bns
1. Grenadier-Battalion,	Manteuffel (37/40)	1
2.	Wedell (1/23)*	1
3.	Ramin (19/25)*	1
4. Field-Infantry-Regiment,	Bornstedt (20), 1st Bn	1
5.	Asseburg (27), 1st Bn	1
6.	Itzenplitz (13)*	2
7.	Meyerinck (26)*	2

8.	Free-Battalion No. 1 Le Noble	1
	No. 3 Kalben	1
10.	No. 4 d'Angellelli	1
11.	Jäger-zu-Fuss	B
12.	Grenadier-Battalion Kremzow (17/22)*	1
13.	Field-Infantry-Regiment,	
	Margraf Karl (19)*	2
14.	II/III Garde (15)*	2
15.	Retzow (Garde-Grenadier) (6)*	1
16.	Kannaker (30)	2
17.	Pannwitz (10)	2
18.	Geist (8)	2
19.	Winterfeldt (1)*	2
20.	Forcade (23)*	2
21.	Alt-Braunschweig (5)*	2
22.	Grenadier-Battalion Hacke (3/6)	1
23.	Schenkendorff (35/36)	1
24.	Unruth (45/48)	1
25.	Kleist (4/16)	1
26.	Diringshofen (21/27)	1
27.	Fusilier-Regiment, Kurssell (37), 1st Bn	1
28.	Field-Infantry-Regiment,	
	Prinz v. Preussen (13), 1st Bn	1
29.	Fusilier-Regiment, Munchow (36), 1st Bn	1
30.	Jung-Braunschweig (39), 1st Bn	1
31.	Alt-Württemberg (46), 1st Bn	1
32.	Grenadier-Battalion, Ostenreich (29/31)	1
33.	Ploetz (Standing Grenadier Bn 6)	1
34.	Field-Infantry-Regiment,	
	Prinz Ferdinand (34), 1st Bn	1
35.	Grenadier-Battalion, Kahlden (SGB) 1	1
36.	Fusilier-Regiment,	
	Prinz Heinrich (35), 1st Bn	1
37.	Field-Infantry-Regiment, Kalckstein (25)	2
38.	Artillery	
		47B

2. *Cavalry*

		Sqns
39.	Hussar-Regiment,	
	No. 2 Zeithen	10
40.	No. 6 Werner	10
41.	No. 3 Warnery	5
42.	No. 7 Seydlitz	5
43.	No. 12 Württemberg	5
44.	No. 7 Seydlitz	5
45.	No. 4 Puttkammer	10
46.	Cuirassier-Regiment,	
	No. 13 Gardes du Corps*	3
47.	No. 10 Gensd'armes*	5
48.	No. 8 Rochow*	5
49.	No. 5 Margraf Friedrich	5
50.	No. 6 Schönaich	5
51.	No. 1 Krochow	5
52.	No. 11 Karabiniers	5
53.	No. 7 Dreisen*	5
54.	Dragoon-Regiment,	
	No. 5 Bayreuth	10
55.	No. 1 Normann	5
56.	No. 4 Czettritz	5
57.	No. 2 Krockow	5
58.	No. 11 Stechow	5
59.	Cuirassier-Regiment, No. 12 Kyau	5
60.	No. 9 Prinz Schönaich	5
61.	No. 4 Gessler	5
		133

3. *Formation*

Vanguard, Cavalry – 39–45
Infantry – 1–11
First Line – 46–50, 12–23, 51–54
Refused wings of central part of line – 24/25, 26/27
Second Line – 55–58, 28–37, 59–61

2. Austrian and Imperial Order of Battle

1. Infantry and Artillery

		Bns
1.	German Infantry-Regiment, (in 1769) No. 1 Kaiser*	2
2.	No. 7 Neipperg	2
3.	No. 12 Botta d'Adorno	2
4.	No. 59 Leopold Daun	2
5.	No. 26 Puebla	2
6.	No. 21 Prinz d'Arenberg	1
7.	No. 10 Alt-Wolfenbüttel	2
8.	Hungarian Infantry-Regiment, No. 37 Josef Esterhazy	2
9.	No. 33 Nicholas Esterhazy	2
10.	German Infantry-Regiment, No. 25 Thürheim	1
11.	No. 49 Kheul	2
12.	No. 13 Moltke	1
13.	No. 8 Sachsen-Hildburghausen	2
14.	No. 2 Erzherzog Karl	2
15.	No. 3 Karl v. Lothringen	1
16.	No. 35 Waldeck	1
17.	No. 11 Wallis	1
18.	No. 15 Pallavicini	1
19.	No. 17 Kollowrat-Krakowsky	2
20.	No. 27 Baden-Durlach	1
21.	No. 4 Hoch v. Deutschmeister	1
22.	Würzburg 'Red' Infantry-Regiment	1
23.	German Infantry-Regiment, No. 36 Josef Browne	1
24.	Hungarian Infantry-Regiment, No. 52 Bethlen	1
25.	German Infantry-Regiment, No. 50 Harsch	2
26.	No. 23 Baden-Baden	1
27.	No. 42 Gaisruck	1
28.	No. 29 Ludwig Wolfenbüttel	2

29.	No. 47 Harrach	2
30.	No. 9 Los Rios	1
31.	No. 16 Königsegg-Rothenfels	1
32.	No. 30 Sachsen-Gotha	1
33.	No. 38 Ligne	1
34.	German Infantry-Regiment, No. 55 Arberg	1
35.	Hungarian Infantry-Regiment, No. 31 Haller	1
36.	German Infantry-Regiment, No. 57 Andlau	1
37.	No. 56 Merci-Argentau	1
38.	No. 45 Heinrich Daun	2
39.	Hungarian Infantry-Regiment, No. 19 Leopold Pallfy	1
40.	No. 31 Haller v. Hallerstein	1
41.	German Infantry-Regiment, No. 46 McGuire	1
42.	No. 48 Luzan	1
43.	No. 44 Clerici	1
44.	Hungarian Infantry-Regiment, No. 32 Forgàch	1
45.	No. 34 Adam Batthyany	1
46.	Hungarian Infantry-Regiment, No. 39 Johann Pallfy	1
47.	Württemberg Infantry-Regiment, Trüchfetz	2
48.	Roeder	2
49.	Prinz Louis	2
50.	Leib	2
51.	Spitznatz	2
52.	Austrian (2) Grenadier-Battalion, No. 1 Pletz	1
53.	No. 2 Georgi	1
54.	No. 3 Klettenberg	1
55.	Bavarian Infantry-Regiment, Kurprinz Brensing	2
56.	Herzog Clemens	2
57.	Leib	2
58.	Morawitzy	2

59.	Minucci	2
60.	Croats	3
61.	Austrian Artillery	
62.	Austrian Artillery-Fusiliers	
63.	Württemberg Artillery	
64.	Bavarian Artillery	
		88

3. *Formation*

1. *Prince Charles of Lorraine*
   Advance Guard – 65, 66, 60, 67–69, 70, 60, 71
   First Line – 72–76, 1–14, 77–81
   Second Line – 82–85, 15–29, 86–89
2. *Herzog v. Arenberg*
   Reserve – 30–33, 34–37
3. *Graf Nadasty*
   First Line – 90, 91, 38–54, 92, 93
   Second Line – 55–59

3. MINDEN

1. Allied Order of Battle

1. *Infantry and Artillery*	*Bns*
1. British Infantry-Regiment, Napier (12th)	1
2. Royal Regiment of Artillery	
3. British Infantry-Regiment, Stewart (37th)	1
4. R. Welch, Fusiliers (23rd)	1
5. Hanoverian Infantry-Regiment, Garde	2
6. British Infantry-Regiment, Kingsley (20th)	1

7.	Brudenell (51st)	1
8.	Holmes (25th)	1
9.	Hanoverian Infantry-Regiment,	
	6A Hardenberg	1
10.	3A Reden	1
11.	2B Scheele	1
12.	4B Stoltzenberg	1
13.	12B Bruncke	1
14.	Hessian Infantry-Regiment, Erbprinz	1
15.	Hanoverian Artillery	1
16.	Hanoverian Infantry-Regiment, 9A Saxe-Gotha	1
17.	Hessian Infantry-Regiment, Toll	1
18.	Schomberg-Lippe-Bückeburg Artillery	1
19.	Hessian Infantry-Regiment, Garde	1
20.	Hanoverian Infantry-Regiment, 7A Wangenheim	1
21.	Hessian Infantry-Regiment,	
	Mansbach	1
22.	Bischhausen	1
23.	Prinz v. Anhalt-Birnberg	1
24.	Brunswick Infantry-Regiment, Behr, 2nd Bn	1
25.	Hessian Grenadier-Battalion	1
26.	Hessian Infantry-Regiment,	
	Prinz Wilhelm	1
27.	Gilsa	1
28.	Brunswick Infantry-Regiment, Imhoff	2
29.	Hessian Infantry-Regiment, Lieb, 1st Bn	1
30.	British Grenadier-Battalion, Maxwell	1
31.	Hanoverian Grenadier-Battalion	1
32.	Hessian Grenadier-Battalion	1
33.		1
34.	Brunswick Grenadier-Battalion	1
35.		1
36.	Hanoverian Infantry-Regiment,	
	2A Spörckern	1
37.	12A Kielmansegge	1
38.	9B Jung-Zastrow	1
39.	6B Halberstadt	1
40.	5A Laffert	1
41.	11B Oberg	1
42.	1A Scheither	1

43.	3B Schulenberg	1
44.	Hessian Artillery	
45.	Schomberg-Lippe-Bückeburg Infantry-Regiment	1
		45

3. *Formation*

1. *Prince Ferdinand of Brunswick*
   1st Column (right), Lord George Sackville – 56–46
   2nd Column, Major Haase, Hanoverian Artillery – 30 guns and howitzers
   3rd Column, LG v. Spörckern – 8–1
   4th Column, MG v. Scheele – 14–9
   5th Column, Oberst Braun – 16, 15
   6th Column, LG v. Wutginau – 23–17
   7th Column, LG v. Imhoff – 29–24
   8th Column, LG Hertzog v. Holstein-Gottorp – 63–57
2. *Wangenheim's Corps* – 44, 45, 34, 33, 32, 30, 31, 43–36, 71–69, 68–64

1. *Infantry and Artillery* — *Bns*

1.	French Infantry-Regiment, No. 1 Picardie	4
2.	No. 84 La Marche Prince	1
3.	No. 10 Belsunce	4
4.	No. 18 Touraine	2
5.	d'Aumont	2
6.	No. 9 Auvergne	4
7.	No. 42 Rouergue	2
8.	No. 74 La Marche Comte	1
9.	No. 71 Tournaisis	1
10.	German Infantry-Regiment, No. 46 Anhalt	4
11.	French Infantry-Regiment, No. 38 Condé	2

12.	No. 85 Enghien	2
13.	No. 19 Aquitaine	2
14.	No. 58 Vastan	2
15.	No. 12 Du Roi	4
16.	No. 2 Champagne	4
17.	Grenadiers de France, No. 40	8
18.	Grenadiers Royaux	8
19.	French Infantry-Regiment, No. 4 Piémont	4
20.	No. 16 Dauphin	2
21.	German Infantry-Regiment, No. 86 Royal-Bavière	2
22.	No. 92 Royal-Deux-Ponts	2
23.	Swiss Infantry-Regiment, No. 55 Waldner	2
24.	Swiss Infantry-Regiment, No. 63 Planta	2
25.	Saxon Infantry-Regiment, Kur Prinzessen	2
26.	Gotha	1
27.	Prinz Anton	1
28.	Brühl	1
29.	Rochow	1
30.	Lubomirsky	1
31.	Clemens	1
32.	Prinz Josef	1
33.	Friederich Augustus	2
34.	Prinz Xavier	2
35.	Prinz Carl	1
36.	Garde	1
		86

2. *Cavalry* — *Sqns*

37.	French Cavalry-Regiment, No. 1 Colonel-Général	3
38.	No. 29 Marcieu	2

39.	No. 40 Vogué	2
40.	No. 21 Condé	2
41.	No. 5 Du Roi	2
42.	No. 50 Henrichemont	2
43.	No. 51 Moustier	2
44.	Roue	2
45.	No. 17 Bourgogne	2
46.	No. 69 Voluntaires Liégois	2
47.	No. 26 Archiac	2
48.	No. Mestre-de-Camp Général	2
49.	No. 41 Fumel	2
50.	No. 49 Espinchal	2
51.	No. 27 Poli	2
52.	No. 8 Royal-Cravattes	2
53.	Surgères	2
54.	No. 10 Royal-Piémont	2
55.	No. 6 Royal-Etranger	2
56.	No. 45 Crussol	2
57.	No. 54 Noailles	2
58.	No. 36 Balincourt	2
59.	Gendarmerie de France	8
60.	French Cavalry-Regiment,	
	No. 12 Royal de Carabiniers	10
61.	No. 3 Commissaire-Général	2
62.	No. 44 Lameth	2
63.	No. 30 Des Salles	2
64.	No. 25 Penthièvre	2
65.	No. 47 Toustain	2
66.	No. 11 Royal-Allemand	2
67.	German Cavalry-Regiment,	
	No. 70 Nassau-Usingen	2
68.	No. 53 Prince de Württemberg	2
69.	Dragoon-Regiment, No. 13 Apchon	4
70.	Voluntaires de Schömberg	3
71.	German Hussar-Regiment,	
	No. 69 Royal-Nassau	4
		90

3. *French Formation*

1. *Marshal Contades*
   First Line – 1, 2, 4, 5, 7, 8, 9, 37–40, 48–51, 11–16
   Second Line – 3, 6, 10, 41–44, 52–54, 25–36
   Third Line – 45–47, 55–58
   Fourth Line – 59, 60
2. *Reserve – Duc de Broglie*
   First Line – 17, 18, 61–63
   Second Line – 19, 20, 23, 24, 65, 65
   Third Line – 21, 22, 66–68
   Fourth Line – 69–71

4. KUNERSDORF

1. **Prussian Order of Battle**

1. *Infantry and Artillery*

		Bns
1.	Grenadier-Battalion, Jung-Billerbeck (5/30)	1
2.	Lubath (7/30)	1
3.	Heyden (19/25)	1
4.	Bornstedt (13/26)	1
5.	Fusilier-Regiment, Bredow (43)	2
6.	Grenadier-Battalion, Schwartz (35/36)	1
7.	Österreich (29/31)	1
8.	Field-Infantry-Regiment, Margraf Carl (19)	2
9.	Hülsen (21)	2
10.	Frinck (12)	2
11.	Fusilier-Regiment, Prinz Heinrich (35)	2
12.	Field-Infantry-Regiment, Knobloch (29)	2
13.	Goltz (24)	2
14.	Schenkendorff (9)	2

15.	Leswitz (31)	2
16.	Kannitz (2)	2
17.	Bevern (7)	2
18.	Fusilier-Regiment, Wied (41)	2
19.	Grenadier-Battalion,	
	Tann (17/21)	1
20.	Lossow (Gar. I/Gar. II)	1
21.	Beyer (11/14)	1
22.	Fusilier-Regiment,	
	Grabow (47)	2
23.	Diericke (49)	2
24.	Garrison-Regiment, Jung-Sydow (5)	2
25.	Fusilier-Regiment,	
	Bülow (46)	2
26.	Dohna (16)	2
27.	Grenadier-Battalion,	
	Bussche (Gar. 6/Gar. 3)	1
28.	Nesse (2/Gar. 2)	1
29.	Field-Infantry-Regiment, Lehwaldt (14)	2
30.	Fusilier-Regiment,	
	Zastrow (38)	2
31.	Braun (37)	2
32.	Hauss (55)	2
33.	Artillery	
		53

2. *Cavalry* *Sqns*

34.	Cuirassier-Regiment,	
	No. 3 Leibregiment zu pferde	5
35.	No. 1 Schlabrendorff	5
36.	No. 7 Horn	5
37.	No. 5 Margraf Friedrich	5
38.	No. 2 Prinz Heinrich	5
39.	Hussar-Regiment, No. 4 Puttkammer	10
40.	Dragoon-Regiment,	
	No. 2 Krochow	5

41.	No. 8 Alt-Platen	5
42.	No. 6 Schorlemer	10
43.	Hussar-Regiment, No. 1 Kleist	10
44.	Free Corps Kleist, Hussars	1
45.	Hussar-Regiment	
	No. 3 Möhring	5
46.	No. 2 Zeithen	6
47.	Cuirassier-Regiment, No. 12 Spaen	5
48.	Dragoon-Regiment,	
	No. 3 Meinecke	5
49.	No. 11 Jung-Platen	5
50.	Hussar-Regiment, No. 8 Belling	5
		97

3. *Formation*

Vanguard – 1–7
First Line – 34, 35, 8–18, 36–38
Second Line – 39–41, 19–28, 42–44
Reserve – 45–48, 29–32, 49, 50

4. KUNERSDORF

2. **Russian Order of Battle**

1. *Infantry and Artillery*

		Bns
1.	Infantry-Regiment,	
	Rostov	2
2.	Apcheron	2
3.	Pskovsk	2
4.	Vologod	2
5.	Grenadier-Regiment,	
	No. 3	2

6.	No. 4	2
7.	Infantry-Regiment,	
	Neva	2
8.	Kazan	2
9.	Grenadier-Regiment, No. 2	2
10.	Infantry-Regiment,	
	Bielozersk	2
11.	Nijenovgorod	2
12.	Novgorod	2
13.	Nizov	2
14.	Azov	2
15.	Perm	2
16.	Grenadier-Regiment, No. 1	2
17.	Infantry-Regiment,	
	Voronetz	2
18.	Narva	2
19.	Vyborg	2
20.	Arkhanguelogorod	2
21.	Tchernikov	2
22.	St Petersburg	2
23.	Kiev	2
24.	Ouglitz	2
25.	Siberia	2
26.	Moscow	2
27.	Troitsk	2
28.	Viatsk	2
29.	Grenadier-Regiment, Observation Corps	2
30.	Musketeer-Regiment,	
	No. 1 Observation Corps	3
31.	No. 3 Observation Corps	3
32.	No. 4 Observation Corps	3
33.	No. 5 Observation Corps	3
34.	Artillery	
		70

2. *Cavalry*

		Sqns
35.	Hussar-Regiment, Slavonisch-Serbsk	5
36.	Venguersk	5
37.	Serbsk	2
38.	Novoserbsk	5
39.	Grouzinski	5
40.	Horse Grenadier-Regiment, Karagopol	3
41.	Riazan	3
42.	St Petersburg	3
43.	Dragoon-Regiment, Tobolsk	3
44.	Arkhanguelogorod	3
45.	Combined Cuirassier-Regiment	3
46.	Cuirassier-Regiment, No. 3	3
47.	Thronfolger	3
48.	Novotroitsk	3
49.	Kiev	3
50.	Thronfolger	1
51.	Don Cossacks (4,500 strong)	
52.	Tchougouev Field Cossack Regiment	
		53

3. *Formation*

Observation Corps – 51, 29–33, 52
First Line – 35–39, 1–13, 40–44
Second Line – 45–47, 14–26, 48–50
Reserve – 27, 28

3. **Austrian Order of Battle**

1. *Infantry and Artillery*		*Bns*
1. Croats		3
2. German Infantry-Regiment,		
	Los Rios	2
3.	Arenberg	2
4.	Leopold Pfallfy	2
5. Free Corps,		
	Loudon	2
6.	Bethlen	2
7. German Infantry-Regiment,		
	Baden-Baden	2
8.	Waldeck	2
9. Artillery		
		17

2. *Cavalry*		*Sqns*
10. Hussar-Regiment,		
	Nadasti	5
11.	Kalnocky	5
12.	Serbia (Russian)	5
13. Dragoon-Regiment,		
	Kolowrat	5
14.	Württemberg	5
15.	Leichtenstein	5
16. Light Dragoon-Regiment, Löwenstein		10
		40

3. *Formation*

Advance Guard – 1
First Line – 10–14, 2–4, 15, 16
Second Line – 5–8

6. QUEBEC

1. British Order of Battle

1. *Infantry and Artillery*

		Bns
1.	British Infantry-Regiment, Otway (35th)	1
2.	Louisberg Grenadiers, (22nd, 40th, 45th)	1
3.	British Infantry-Regiment, Bragge (28th)	1
4.	Kennedy (43rd)	1
5.	Lascelle (47th)	1
6.	Fraser's Highlanders (78th)	1
7.	Anstruther (58th)	1
8.	Amherst (15th)	1
9.	Webb (48th)	1
10.	Royal American (60th) 2nd Bn	1
11.	3rd Bn	1
12.	Light Infantry Battalion, Howe	1
13.	Rangers	
14.	Marines	
15.	Seamen	
16.	Artillery	
		12

2. *Formation*

First Line – 3 (1 Pln), 1–7

Second Line – 8
Third Line – 9, 10
Fourth Line – 11, 12
Reserve – 7 (2 Coys)

2. French Order of Battle

1. *Infantry and Artillery*

		Bns
1.	Canadian Militia Quebec Colony	1
2.	French Infantry-Regiment,	
	No. 72 Béarn	2
3.	No. 34 La Sarre	2
4.	No. 68 Guyenne	2
5.	No. 53 Languedoc	2
6.	No. 57 Royal Roussillon	2
7.	Canadian Militia,	
	Montreal Colony and Three Rivers	1
8.	Artillery	
9.	Indians	
		12

2. *Formation* – 1–7

8. WARBURG

1. Allied Order of Battle

1. *Infantry and Artillery*

		Bns
1.	British Grenadier-Battalion,	
	Daulhatt	1
2.	Maxwell	1

3.	Hanoverian Grenadier-Battalion,	
	No. 1 Genso	1
4.	No. 3 Bock	1
5.	No. 2 Wierfebe	1
6.	Hanoverian Infantry-Regiment,	
	1A Scheither	1
7.	12A Estorff	1
8.	10A Post	1
9.	British Infantry-Regiment,	
	Keith's Highlanders (87th)	1
10.	Campbell's Highlanders (88th)	1
11.	Hanoverian Infantry-Regiment,	
	8B Block	1
12.	10B Monroy (1st New Bn)	1
13.	Hessian Infantry-Regiment, Garde	2
14.	Brunswick Grenadier-Battalion,	
	Wittorf	1
15.	Redecker	1
16.		1
17.	Hessian Garrison Grenadier-Battalion	1
18.	Hessian Infantry-Regiment,	
	Toll	2
19.	Pappenheim	1
20.	Mirbach	1
21.	Rüchersfeld	1
22.	Hanoverian Light Troops, Légion Britannique	5
23.	British Artillery	
24.	Hanoverian Artillery	
25.	Hessian Artillery	
		28

2. *Cavalry* — *Sqns*

26.	Hanoverian Dragoon-Regiment, Breitenbach	4
27.	British 1st Royal Dragoons, (Conway)	2
28.	Hessian Cavalry-Regiment, Einsiedel	2
29.	British 7th Dragoon-Regiment, (Cope)	2

30.	Hessian Cavalry-Regiment, Pruschenk	2
31.	Hanoverian Cavalry-Regiment, No. 3 Bremer	2
32.	Hanoverian Dragoon-Regiment, No. 4 Bock	4
33.	No. 3 Reden	4
34.	British 1st Kings Dragoon Guards (Bland)	3
35.	3rd Dragoon Guards (Howard)	2
36.	2nd Dragoon Guards	2
37.	Royal Regiment of Horse Guards (Blue)	3
38.	7th Dragoon Guards	2
39.	Carabiniers	2
40.	2nd Royal North British Dragoons (Campbell)	2
41.	10th Dragoons (Mordaunt)	2
42.	6th Inniskilling Dragoons (Cholmondeley)	2
43.	11th Dragoons (Ancram)	2
44.	Hanoverian Light Troops, Légion Britanniques	5
		49

3. *Formation*

1. *Erbprinz*
   Right Column – 1–12, 26–30
   Left Column – 13–21, 31–33
   Detached – 22, 43
2. *Granby*
   Front Line – 34–39
   Second Line – 40–43

2. French Order of Battle

1. *Infantry and Artillery* — *Bns*

1.	French Infantry-Regiment, No. 7 Latour du Pin	4
2.	No. 18 Touraine	2

3.	No. 85 Enghien	2
4.	Swiss Infantry-Regiment,	
	No. 63 Planta	2
5.	No. 78 Courten	2
6.	No. 90 Lochmann	2
7.	No. 51 Reding	2
8.	French Infantry-Regiment,	
	No. 28 La Couronne	2
9.	d'Aumont	2
10.	No. 8 Bourbonnois	4
11.	No. 42 Rouergue	2
12.	No. 64 Rochefort	2
13.	No. 47 Royal Artillery	
		28

2. *Cavalry*		*Sqns*
14.	French Cavalry-Regiment,	
	No. 6 Royal-Étranger	2
15.	No. 38 St Aldegonde	2
16	No. 26 Archiac	2
17.	No. 14 La Reine	2
18.	No. 45 Crussol	2
19.	No. 36 Balincourt	2
20.	No. 22 Bourbon	2
21.	No. 68 Montcalm	2
22.	No. 35 Beauvilliers	2
23.	No. 10 Royal-Piémont	2
24.	No. 58 Escars	2
25.	No. 49 Espinchal	2
26.	Dragoon-Regiment,	
	No. 14 Thiange	4
27.	No. 4 Royal	4
28.	French Light Troops, Chasseurs de Fischer	4
		36

3. *Formation*

First Line – 1–3, 14–27, 4–10
Second Line – 11, 12
Detached – 28

BIBLIOGRAPHY

Andolenko, General, *Histoire de l'Armée Russe*, Flammarion, Paris, 1967.

Annual Register, 1776.

Atkinson, C. T., *Regimental History of The Royal Hampshire Regiment*, 2 *Vols*, Maclehose, Glasgow, 1950.

Breville, J. O. de (Job), *Tenue des Troupes de France*, 4 Vols, Combet, Paris, 1903.

& Bouchot, Henri, *L'Epopée du Costume Militaire Français*, May, Paris, ND.

Brunon, Raoul and Jean, *Les Arts Militaires Français au XVIIIe Siècle*, Aix-en-Provence, 1962.

Carman, W. Y., *British Military Uniforms from Contemporary Pictures*, Leonard Hill, London, 1969.

Cary, A. D. L., OBE, & McCance, Captain S., *Regimental Records of the Royal Welch Fusiliers*, Foster Groom, London, 1921.

Chichester, H. M., & Burges-Short, G., *The Records and Badges of every Corps in the British Army*, Gale & Polden, London, 1899.

Cole, Lieut.-Colonel Howard N., OBE, TD., *Minden* 1759, Charles Knight, London, 1972.

Colin, Commandant, *L'Infânterie au XVIIIe Siècle, La Tactique*, Berger-Levrault, Paris, 1907.

Duffy, Christopher, *The Army of Frederick The Great*, David & Charles, London, 1974.

The Wild Goose and The Eagle, A Life of Marshal von Browne, Chatto & Windus, London, 1964.

Fontblanque, E. B. de, *Political and Military Episodes . . . derived from the Life and Correspondence of the Right Hon. John Burgoyne*, Macmillan, London, 1876.

Fortescue, Hon. J. W., *A History of the British Army*, Vol. II, Macmillan, London, 1911.

Fuller, Colonel J. F. C., *British Light Infantry in the Eighteenth Century*, Hutchinson, London, 1952.

Fürstliches jagen, (The Tischbein paintings in Schloss Fasanerie), Insel, Frankfurt am Main, ND.

Gazette des Uniformes.

Haswell Millar, A. E., & Dawnay, N. P., *Military Drawings and Paintings in the Collection of Her Majesty the Queen*, 2 Vols, Phaidon, London, 1970.

Heere und Tradition.
Higgins, Captain R. T., *The Records of the King's Own Borderers or Old Edinburgh Regiment*, Chapman & Hall, London, 1925.
History Today.
Jarret, Dudley, *British Naval Dress*, Dent, London, 1960.
Journal of the Society for Army Historical Research.
Knötel, Herbert, & Sieg, Herbert, *Handbuch der Uniformkunde*, Schultz, Hamburg, 1937.
Knötel, R., *Uniformenkunde*, 18 Vols, Rathenow, 1890–1921.
& Röchling, Carl, *Der Alte Fritz*, Kittel, 1895.
Kornath, Friedrich, (ed.), *Dessins des Uniformes des Troupes I.II et R.R. de l'année* 1962, facsimile edition of the Albertina MS, Tusch, Vienna, 1973.
Lawson, Cecil C. P., *A History of the Uniforms of the British Army*, Vol. 2, L. Peter Davies, London, 1941.
Leach, Douglas Edward, *Arms for Empire, A Military History of the British Colonies in North America*, 1607–1763, Macmillan, New York, 1973.
Lezius, Martin & Knötel Herbert, *Deutsche Uniformen das zeitalter Friedrichs des Grossen*, Sturm-Zigaretten, Dresden, 1932.
Macdonald, Captain R. J., *The History of the Dress of the Royal Regiment of Artillery*, Sotheran, London, 1899.
Mangerand, J., *Armement et Equipement de l'Infânterie Française*, L'Editions Militaires Illustrés, Paris, 1945.
Maude, Lieut.-Colonel F. N., *Cavalry: Its Past and Future*, Clowes, London, 1903.
May, Commander W. E., RN, *The Dress of Naval Officers*, HMSO, London, 1966.
Military Collector & Historian.
Millar, Oliver, *The Later Georgian Pictures in the Collection of Her Majesty the Queen*, 2 Vols, Phaidon, 1969.
Mollo, John, *Military Fashion*, Barrie & Jenkins, London, 1972.
Mouillard, L., *Les Regiments sous Louis XV*, Dumaine, Paris, 1882.
Neptunia.
Ottenfeld, R. von, & Teuber, O., *Die Osterreichische Armée*, 1700–1867, Vienna, 1895.
Rambaud, Alfred, *Russes et Prussiens, Guerre de Sept Ans*, Berger-Levrault, Paris, 1895.
Revol, Colonel J., *Histoire de l'Armée Française*, Larousse, Paris, 1929.

Reynolds, P. W., *Military Costume of the 18th and 19th Centuries*, unpublished MS, c.1891.
Rousselot, L., *L'Armée Française, Ses Uniformes, Son Armement, Son Equipement*, privately published series of plates, current.
Sabretache, Carnet de la.
Savory, Lieut.-General Sir Reginald, KCIE, CB, DSO, MC, *His Britannic Majesty's Army in Germany during the Seven Years War*, Oxford University Press, 1966.
Schirmer, Friedrich, *Das Celler Soldatenbuch*, Pohl, Celle, 1937.
Nec Aspera Terrent!, Hanover, 1929.
Seele, W. (ed.), *Die Schlacht bei Minden* 1759, Bruns, Minden, 1959.
Die Siebenjahrige Krieg 1756–1763, *Herausgeben vom Grossen Generalstabe Kriegsgeschichtliche Abteilung II*, Berlin, 1901, etc.
Telfer-Dunbar, J., *History of Highland Dress*, Oliver & Boyd, Edinburgh and London, 1962.
Tradition.
Viskovatov, A. V., *Historical Description of the Uniforms and Weapons* of The Russian Army, 30 Vols, St Petersburg, 1844–1856.
Verillon, M., *Les Trophées de la France*, Leroy, Paris, 1907.
Vert et Rouge.
Warnery, Major-General, *Remarks on Cavalry*, Edgerton, London, 1798.
Windham, W., & Townshend, Lord G. Viscount, *A Plan of Discipline composed for the use of the Militia of the County of Norfolk, London*, 1759.
Zweguintzov, W., *L'Armée Russe, Vol.* 1, Paris, 1967.